Doug Pratt's Modeling Guides

D1503126

Hey Kid!...
Ya Wanna Build
A Model Airplane?

A Beginning Guide
to Building the Peck R.O.G.

Bill Warner

TAB TAB BOOKS
Blue Ridge Summit, PA

FIRST EDITION
FIRST PRINTING

© 1991 by **TAB Books**.
TAB Books is a division of McGraw-Hill, Inc.

Based on articles from *Model Builder Magazine*.

Library of Congress Cataloging-in-Publication Data

Warner, Bill.
    Hey kid!—Ya wanna build a model airplane? : a beginning guide to
building the peck R.O.G. / by Bill Warner.
        p.   cm.
    Includes index.
    ISBN 0-8306-1040-5 (pb)
    1. Airplanes—Models.   I. Title.
TL770.W285   1991
629.133′134—dc20
                                                          91-20593
                                                          CIP

TAB Books offers software for sale. For information and a catalog, please contact
TAB Software Department, Blue Ridge Summit, PA 17294-0850.

Acquisitions Editor: Jeff Worsinger
Book Editor: April D. Nolan
Production: Katherine G. Brown
Book Design: Jaclyn J. Boone

Contents

Acknowledgments

*T*he author would like to thank Bill Northrop of *Model Builder Magazine* for his support of free-flight model building and for inspiring this series. His kind permission to use the material from the "Hey Kid" series has made this book possible.

Special thanks are also in order to Bob and Sandy Peck of Peck Polymers, who have given invaluable assistance to beginning modelers and have helped immeasurably in making the excellent models we have chosen to feature in this work easily available.

An immense debt of gratitude is due to the unsung and innumerable modelers who have taken the time to help beginners such as we have all been, to the model clubs such as the Flying Aces who have kept free flight alive and well, and to the many newsletter editors who keep us well-supplied with plans, hints, and tips. Without their aid, most of us would have given up and missed out on the world's greatest hobby: building and flying free-flight model airplanes!

To the memory of Bob Peck
1934—1991

Introduction

*T*his is the first of a three-part series of books incorporating the popular "Hey Kid!" series of beginners' articles which originally appeared in *Model Builder* magazine. The five chapters in this book correspond with the first five articles of the 14-part magazine series. Two more books are planned to cover the remainder of the episodes, and with the basics of aeromodeling you will learn in this book, you can apply your knowledge to the more complex models and techniques discussed in volumes two and three.

This book is designed to give the beginner a complete, step-by-step, tested approach to success in model airplanes. The projects include modifying the Sleek Streek model—a version of the simple, balsawood planes you often find in drugstores—so that it will actually fly. You will also learn how to build and fly the Peck R.O.G. (Rise-Off-Ground) model, one of the easiest beginner models to build & fly.

Note: The Sleek Streek, Peck R.O.G., model knife, glue, winder, and other materials used in this book may be obtained from Peck Polymers (address in the appendix).

Both of these projects have been selected with an eye to easy availability by mail order, and they are readily adaptable to youth groups.

Although this book was designed with beginners in mind, much of the mail received during the run of the series in *Model Builder* was from older modelers who had given up the hobby for many years and found new inspiration to get restarted in the hobby from the articles. The models chosen for the series as a whole were picked for their contribution to the learning of new concepts and techniques and for their excellent flyability.

By the time you have finished this book, you'll be able to make a model capable of flying on its own for possibly quite a long time. In addition, what you have learned will enable you to branch out to any of thousands of rubber-powered models available in kits and plans—enough to last you for many lifetimes. You might also find you want to compete with your models in national contests. Whatever the case, I hope you'll stay with the hobby, and that you will enjoy making model aircraft as much as I do.

Chapter 1

Modifying the Sleek Streek

*M*y first miniature plane was a silk-covered, wind-up, made-in-Japan model that refused to fly, but which made a great fire when I set light to it. My grandmother had bought it for me with the assurance that it was "ready to fly!" Well, it wasn't, and not much has changed since that fateful day in 1937. Millions of kids buy little balsa R.O.G. (Rise-Off-Ground) "Sleek Streek"-type models and later blame themselves when the models crash again and again. Soon, these kids move on to other things, leaving a trail of broken balsa bits behind them. So what's wrong? They need someone to help them over the rough spots and give them a basic course in making them and making them fly! That's what we're going to try to do for you kids from 9 to 90—share fifty years' worth of painfully gotten knowledge. If your head is in the clouds and you can't seem to get your butt off the ground, read on!

First of all, those little R.O.G.s you buy at the corner drugstore are seldom more than a toy item designed by a good modeler and then messed up by companies who cut corners to make a buck. After such models are relieved of some excess weight, given a decent motor, hopped-up here and there, they fly pretty darned well. My test pilot, Kris Samonas (age 10) managed to put a stretched version of the Sleek Streek on top of a two-story building after a flight of over a minute from the beach parking lot in Santa Monica while we were taking the pictures for this book. For a model that is properly made and flown, this was not an unusual flight. When there are rising air currents or *thermals*, flights of several minutes happen all the time.

I chose the Sleek Streek for beginners to use for just this reason. It flies well, and it can teach you a lot about how any model flies. It can be repaired easier than tissue-covered models, and if you lose one or two in a

Author Bill Warner holding his Free-Flight Stinson SR-7 Reliant, with which he won first place at the US Free-Flight Championships in Taft, California, in 1987.

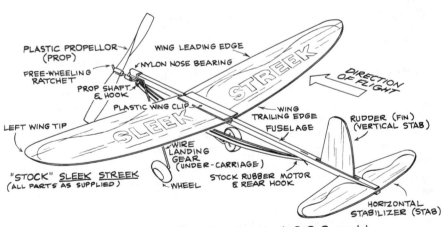

General layout of the Sleek Streek, a simple R.O.G. model.

morning's flying, you aren't out a great deal of money for the fun you had. In my many years of teaching classes in modeling and sponsoring model clubs, I have seen a couple thousand Sleek Streeks made and flown in various degrees of modification with more success than any other model I can think of (except the Midwest X-18 designed by Frank Zaic—see the appendix).

Materials and tools needed to hop up the Sleek Streek: pliers, knife, sanding block, thread, pin, and cellulose cement.

Peck Polymers has agreed to add Sleek Streeks to their line of model kits, plans, and supplies for those of you who can't find them at your neighborhood drugstore. They will send you a sheet with all the models and materials mentioned in this book from rubber strip to "peanut-scale" models. I decided to use Peck models in this attempt to get you started for the simple reason that many of you do not live near a good hobby shop. And besides that, they fly great! Send a self-addressed stamped envelope for a price list of all the models and stuff you'll be using in later chapters. See the appendix for the address of Peck Polymers and those of other suppliers you might want to contact.

GETTING MORE LIFT AND LESS WEIGHT

The wing on a model is the most important part. The wings of many Almost-Ready-to-Fly (ARF) models are warped, heavy, or poorly designed, which is discouraging for many kids. The wing gives the model its *lift*, needed to overcome the weight of the model and get it into the air. The *cambered* (curved) top surface of the wing speeds up the flow of air on the top of the wing only, lowering the air pressure there. The bottom surface of the Sleek Streek wing is *undercambered* (hollow curved) which helps force the air flowing rearward down, which in turn creates higher pressure under the wing.

The *leading edge* of the wing, or the front edge that meets the wind, is also called the LE and needs to be kept a little higher than the *trailing edge* or TE, the rear edge of the wing.

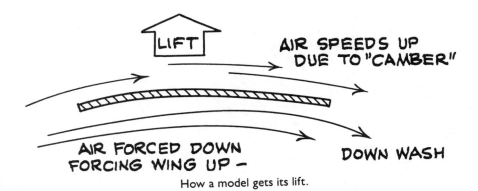

How a model gets its lift.

About 200 years ago, a Swiss scientist named Bernoulli found that when you moved a fluid over a surface, it lowered the pressure on that surface. Air works that way, too. The combination of a cambered upper wing surface along with an *alpha* or *angle of attack* created by keeping the LE higher than the TE in flight gives lift. The undercamber helps, too. If you keep your wing nicely cambered, it will have more lift than a flat wing. Can you guess why the first thing you should throw away is the heavy red plastic clip that flattens the part of the wing nearest the model's *fuselage* or body? First, it saves weight, and second, it lets you get camber into the whole wing for more lift.

Continuing our crusade against weight, take a pair of pliers and straighten out the curved part of the staple that holds the rear of the motor. Shove it out and throw it away: A bent pin will work much better. Now with your thumbnail, try and dent the balsa fuselage under the slot where the "stab" goes. (You'll cut this away soon anyway.) If it is easy to dent, only cut off the lower one-third of the fuselage material to lighten it

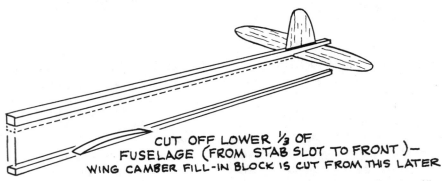

Cut off the lower one-third of the fuselage (from stab slot to front). Wing camber fill-in block is cut from this later.

The upper model is the sport version R.O.G. (Rise-Off-Ground). The lower version is designed for high performance, with no landing gear and a stretched fuselage.

up. If it is hard to dent, you can probably cut off the lower half of the fuse-lage and still have enough strength. You have to balance your need for lightness with your need for strength.

Caution: Save part of what you cut off for the fill-in camber block, and also leave the top part of the stab slot so you can glue the stab to it and keep the angle correct, even though the bottom half of the slot has been cut away. This angle is terribly important because it controls the angle of attack of the wing. Too little or no angle, and you don't get much lift!

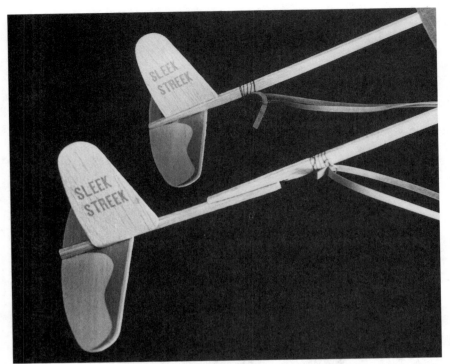

The fuselage may be extended as shown for more stable flights. Use material cut from the underside of the fuselage for the tail boom.

Cut a chunk out of the nylon nose bearing and glue what remains to the fuselage with cellulose glue such as Testors "green tube" (fast-drying) or Duco cement. (I advise against the use of instant glues, having had three good friends wind up in the intensive care units of their local hospitals with that stuff in their eyes.) Bind with a few winds of thread and rub glue into the thread to hold it. Make a bent pin for the rear rubber motor hook and glue and bind it the same way. Now you have clearance and more for your longer, weaker, and longer-running rubber motor. Use an 18″ loop of 1/8″ Sig rubber or a 3/32″ FAI rubber, which you can get through mail order if your hobby shop is out of them.

You can lighten your model even more by sanding the wing and tail parts, especially if they are of heavy balsa. Sometimes, the balsa in the package is so heavy that it requires replacement. If you use lighter balsa, camber it by wetting the top surface a bit and then gluing a couple of extra fill-in camber blocks under the wing between the third and fourth letters of "Sleek Streek" on each side. The artist for the book, Jim Kaman, made some up with wings that were pretty flat, and he says they flew great for him. I've noticed that cambered wings have less tendency to warp.

Note how the wing position is aft (rearward on stretched-fuselage version, on top). This is to move the center of lift back over the new balance point (center of gravity).

-STOCK-

REMOVE

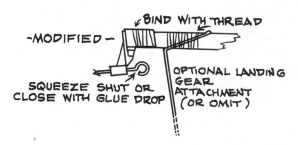

BIND WITH THREAD

-MODIFIED-

SQUEEZE SHUT OR
CLOSE WITH GLUE DROP

OPTIONAL LANDING
GEAR
ATTACHMENT
(OR OMIT)

REPLACEMENT PROP SHAFT-
1/32 MUSIC WIRE - ADD BEAD OR SMALL
BRASS WASHERS

Remove a chunk of the nylon nose bearing and use a bent pin for the rear rubber motor hook.

For the sport version, glue the stab under the rear of the fuselage where the slot was. For the stretched version, make a tail boom about 4″ long out of the scrap you cut from the fuselage, gluing the front of the boom to the slot angle. Be sure to wipe off all extra glue. A thin coat dries

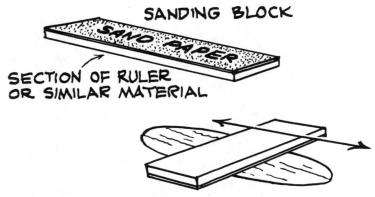

Always sand with the grain, and keep workpiece on a flat surface.

Note the stab angle: Air hits on top, forcing the tail down and the nose up.

a hundred times faster than a thick one. Gobs of glue stay gooey inside. "Double-gluing" is a good plan. Put on some glue, put the parts together, wipe off the glue that squeezes out of the joint, and then take them apart and let them dry. This lets the glue soak into the wood and fill up the pores. Then, glue it again the same way, but this time do not take them apart. Sometimes it helps to wait about 20 seconds until the glue gets "tacky," as it will grab better than wet glue.

You can make a sport version with landing gear—essentially a stretched, high-performance version without undercarriage. The longer-fuselage, gearless version has less weight and drag. The Rise-Off-Ground

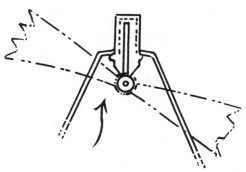

If you keep the wire landing gear, rebend it with pliers as shown to clear the motor.

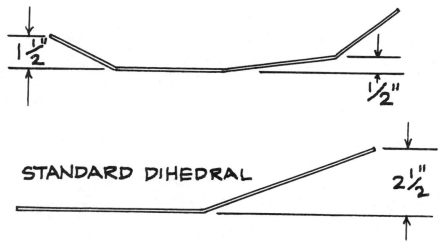

STANDARD DIHEDRAL

Dihedral angles for standard dihedral and polyhedral.

version is fun and performs almost as well. If you make the sport version, be sure to bind the landing gear wire with thread and use enough glue so that it will not move around—it should be "shouldered" out to keep the longer motor from hitting it.

The wing, being the heart of the model, deserves special attention. Using your sanding block—which can be as simple as a bit of yard stick with some 100-grit garnet paper glued on it—sand the wing thinner if it is hard balsa. Finish it up with 400-grit wet-or-dry paper (your hardware will know what that is) until it is "smooth as a moth's nose." Some modelers believe that the first 3/4″ behind the leading edge on top (printed red) should be left rough, because it helps stir up the air on top of the wing and thereby keeps it from peeling off about halfway back, giving more lift. This is called keeping the *boundary layer* (air that sticks to the wing) *turbulent* or stirred up.

The most difficult part of the model is the wing joints, so pay close attention. As you look at the model from the front, you will notice that the stock model has *dihedral*—that is, the tips of the wings are higher than the center or *roots* of the wings. This is an important feature in a model that flies all by itself. In a turn, the wing on the outside of the turn travels a little faster and therefore makes too much lift for the plane to stay level. The dihedral automatically lifts the inside wing that would otherwise drop in a turn. As the inside wing drops more and more toward the horizontal position, it gets more and more lift during the turn. At the same time, the outside-of-turn wing has its tip getting higher and higher, which makes it lose lift (If it went all the way up vertically, it would have no more lift at all, right?). The dihedral effect also helps if your model gets upset by a gust of wind.

This tendency to level the wings side-to-side is called *lateral stability*. You can put in the same dihedral as the wing had in its stock form, or you can add *polyhedral* (more than one dihedral break), which allows more lift in the center section, where you need efficiency.

To sand the correct angle at the root while keeping the camber in the wing, you will need to block up the camber and sand carefully at the edge of your workbench with the sanding block held at the angle shown in the sketch.

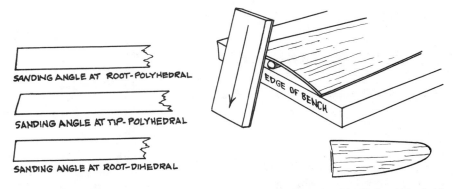

Block up under the side of the wing to keep camber. The sanding block should be held at an angle. After sanding, the flattened wing should slightly curve in.

When you finish sanding, the root will be slightly curved inward at the center if you have done it right. Sand a little and try the wing parts together often for a correct fit-up.

Polyhedral wings are not much different. Cut the wing tips off about 2 1/2 " in from the end and sand the same way to maintain the camber. Change the angle on the sanding block so you have about a 1 1/2 " rise from the center panel to the wing tip, and about 1/2 " rise from one center panel to the other. (See the dihedral sketch.)

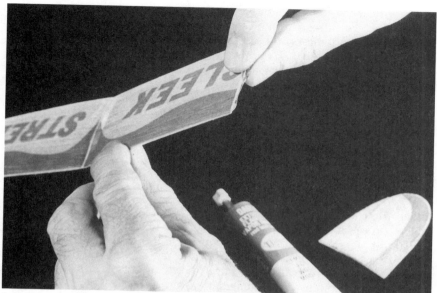

A "V" between the thumb and first finger leaves a thin line of cement on the edge of the joint and wipes off the extra.

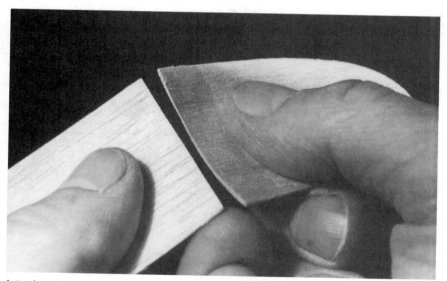

Join the wing parts like this to keep the wing cambered. Do not flatten it out on the table, or it will take the camber out.

Spread a little glue on the edge of each part that fits together, using the double-gluing technique. Use great care not to flatten out the camber. Before the wing is dry, check the measurements for dihedral. Take it apart and do it over if it is not pretty close, within about 1/4".

Set the wing on its leading edge to dry so it will not droop. When completely dry, spread a thin skin of glue (no thicker than a sheet of paper) on the top and bottom of each joint. Keep the width of this glue coating to no more than 1/4" or you will wind up with a warp in your wing. Cellulose glue shrinks when it dries and should not be smeared carelessly about.

While this is drying, make a little block to fill in the space between the wing and the fuselage. You can use the balsa you trimmed off the fuselage earlier for the block. Glue it under the center of the wing, let it dry, and then level it off with your sanding block so that it does not extend down below the LE or the TE.

BALANCING THE MODEL

Attach the rubber motor to the fuselage along with everything else that will be on the model when it flies. Balance it like a teeter-totter on your finger and mark where it balances. Now glue the wing centered up with the red part facing forward right over the mark. Attaching the wing is always left until last because you have to have the fuselage finished, with the rubber and landing gear installed, before you know where your *center of gravity* (CG) will be. The *center of lift*, a point about halfway between the LE and the TE, must be located just about over the CG or the model will be out of balance. Full-size aircraft have to pay close attention to this, too. If the wing is too far forward, the tail will be too long and heavy and will hang down, giving your wing too much angle of attack—which will

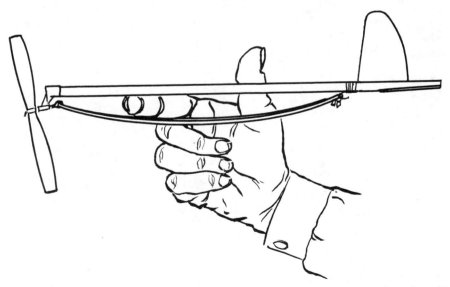

Find the balance point with the tail/prop assembly (and landing gear, if used), and mark it carefully.

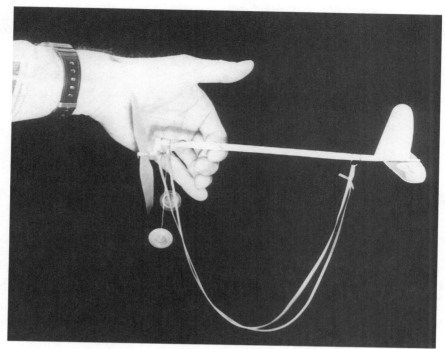

Teeter-totter balance the entire model, minus wing but with rubber, to see where the wing will go.

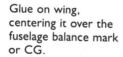

Glue on wing, centering it over the fuselage balance mark or CG.

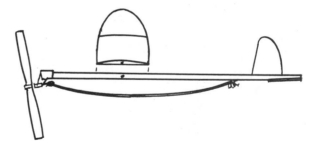

ultimately result in a loop or a stall. If the wing is too far rearward or *aft* the nose will be too long and heavy, reducing the alpha as it droops and giving you a dive into the ground—or, at best, a too-fast descent. The Sleek Streek as it came from the package had the little red clip (that we threw away) to let you slide the wing forward or aft, moving the center of lift. Now the only control you have is where you put the wing. Double check it to see if the model, with the rubber attached, balances under the wing.

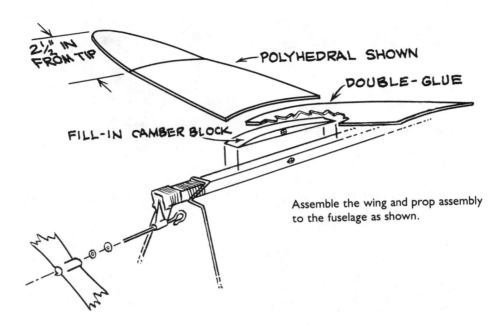

2½" IN FROM TIP

←POLYHEDRAL SHOWN

DOUBLE-GLUE

FILL-IN CAMBER BLOCK

Assemble the wing and prop assembly to the fuselage as shown.

CHECK ALL ALIGNMENTS

Before the wing dries solid, hold the model out at arm's length and close one eye. Sight from front to back to make sure the stab is level with the wings. The model will want to turn towards the side of the stab that is "high," so get it level with the wing now.

Ready to fly? Let's read the flying instructions in chapter 2 first!

Pre-flighting
the Sleek Streek

*T*here's an old saying that any idiot can build a model, but it takes a genius to get one to fly. I think that idiots probably go out and buy their models ready-made, and that geniuses probably need a little help getting their first model into the air just like the rest of us. Of the first fifty models that I made in my life, only one flew—and that one flew purely by accident, I'm sure, because I had no idea what made them fly! That one model, though, made all the rest worthwhile as it puttered off over the daisies with its propeller spinning gaily. What a thrill! Unfortunately, because I didn't know what I had done right, it was ten years before I got another one to fly.

After spending some years in the Marines, going to college, and doing other things besides making models, I settled down to teaching junior high school in Los Angeles. The first thing they asked me was, "Can you sponsor any clubs?" Remembering my younger days with models, I volunteered to start a model airplane club. I stopped by my local hobby shop, which luckily for me was run by an old-time modeler who could help me out. Naturally, I picked a model that would give me as much trouble as I had ever had before—a three-foot-wingspan, gas-powered model that I managed to build easily. Getting it to fly was another matter.

I took that model to the flying field every day for six weeks with no success. The engine gave me fits with plugged fuel passages, dead batteries, weak glow heads, blown gaskets, and a host of other fun things. When I finally did get the engine sorted out, the crashes began! Crash and repair, crash and repair. Every flight was fast, short, and discouraging.

Then one night as I was repairing the wing while working my second job (most teachers have them!), I got into a conversation about planes

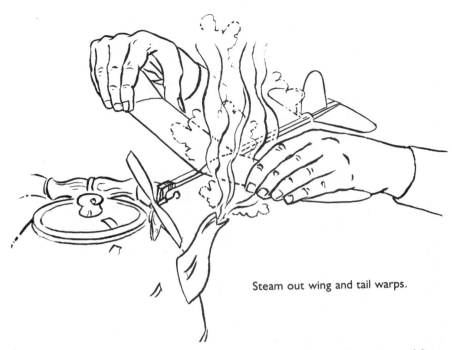

Steam out wing and tail warps.

with one of the chaps who happened by. I told him what my problem was, and he asked me if I had taken the warps out of the wing. "Warps?" I said. "What are warps?"

We heated the wing over a hot plate and twisted it in the opposite direction from the way it had twisted itself and checked it on the flat top of a desk. On my way home that night, it flew so well I lost it out of sight. "Full-tank" Warner had never dreamed such a little thing would make such a difference!

Well, that guy sure knew his stuff. His next advice to me was to get a Sleek Streek and bring it to work. He made up some little tabs from the gummed flap of an envelope and stuck them on the trailing edge of the wings, the rudder, and the stab. After taking the warps out, we flew it inside the office space. He showed me how to make adjustments to keep it from hitting the ceiling or walls. By the time the night was over, I had learned more about flying than I had in the previous twenty years! Since then, having had the good fortune to live in Los Angeles and fly models with some of the best, I have learned a great deal more, but that session with the Sleek Streek was the turning point. If you pay close attention, maybe I can pass on some of these "secrets" to you so that you too can start losing your models instead of just smashing them up!

SOME BASICS

After making your model just right from the instructions in chapter 1, you'll want to stop and think of why each part of the airplane is there.

Nothing is for decoration; even the printing on the wing is to make sure you put it on facing the right direction, so re-check each part.

The fuselage not only holds the motor, but also keeps the wing and the tail in just the right position. The rudder is on straight to keep the plane flying straight ahead. The stabilizer is positioned with the leading edge lower than the trailing edge so that the airflow (which is called the *relative wind*, coming from the front) "sees" the top side of the stab and not the bottom. This allows the air to hit the top side and push on it a bit, creating higher pressure and forcing the tail to go downward a little until it is flat in the airflow. Guess what this does to the position of the fuselage with the wing glued to it? Yep! When the tail is forced down, the nose goes up.

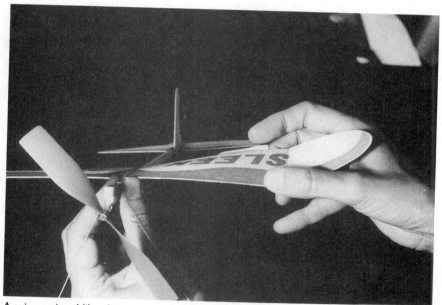

A wing twisted like this is said to have "wash-out." In flight, the airflow will push this wing downward, in this case, rolling the model to the left.

Are you still with me? Now, guess which side of the wing the relative wind "sees"? The underside of the wing is now being hit by the wind and the higher pressure created, and this forces the wing up, which is exactly what you want! Can you see why the stab must be set as it is?

We say the stab has *negative incidence*. Some people prefer to set the stab flat and put *positive incidence* in the wing (leading edge higher than the trailing edge). The important thing is to make sure the wing is going to have a *positive angle of attack* (leading edge higher) as the model flies through the air! Full-size airplanes are no different.

The *center of gravity* (CG) which is often shown on a model plan as a little circle with a cross in it with two of the sections filled in) is located

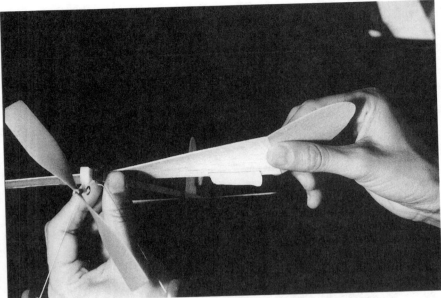

A wing twisted like this is said to have "wash-in." In flight, the airflow will push this wing up and roll the model to the right.

under your wing. If the lift of the model is not over the center of gravity, the model will not want to fly level. If you don't believe me, add some lead to the nose and see if the model doesn't dive in. Add some to the tail and see if the model doesn't swoop upward just the same as if you'd built in too much negative incidence in the stab. You have to balance the weight and the aerodynamic forces (the push-and-pull of the air on various parts of the plane).

On close inspection, you will notice that the propeller shaft is pointed downward toward the front a few degrees. When the plane starts on its flight, the rubber is wound tight, and puts out a great burst of power which gradually runs down as the motor unwinds, making the model fly much faster in the first few seconds of flight. Adding this built-in downthrust to the propeller pulls the front of the model down during the initial power burst when the faster-flying wing is making more lift than you need. If you did not have it, the plane would loop or *stall* (nose comes up, then drops like a roller-coaster).

If you have everything put together exactly right, the model will fly. But what if it is almost but not quite right? That's where *trimming*, or making small adjustments comes into the picture.

PRE-FLIGHT INVENTORY

I once spent a half-hour taking all the warps out of the wing and stab of a six-foot gas job only to find, when I got to the flying field, that I had left the fuselage home. Unless you have a very good memory, you might want to make up a checklist of things to do before you ever go out the hangar door.

RUBBER MOTORS

For the Sleek Streek, you have to have a rubber motor, and the one that comes with it is not very good. You can link a bunch of small rubber bands together to get lower power, but the best thing to do is to get some 3/32″ or 1/8″ flat rubber and make up some loops for motors.

The longer the loop is, the less power it puts out. The shorter a motor is, the more power it puts out. However, a long motor gives a longer prop run, and a short motor gives a short prop run. I suggest you start by making up three motors, one a 10″ long loop, one a 13″ loop, and the other a 16″ loop. Another thing to remember is that a motor made from 3/32″ flat rubber is going to give less power than one made from 1/8″ wide rubber.

Brands differ in performance, too. Some rubber puts out less power, but seldom breaks, whereas other rubber might do just the opposite. Old rubber breaks easily, as does rubber left out in the sun. To make matters worse, the same brand of rubber might vary with different batches! You will notice that FAI rubber (available from FAI, Peck, and Hannan's Runway) is usually thicker and heavier than SIG contest rubber of the same width. The SIG will put out less power but run longer for the same length and width motor—and, of course, it weighs less. Getting an assortment of rubber brands and sizes will make experimenting to find just the right motor easier.

Some modelers prefer to make two loops of 1/16″ rubber rather than one loop of 1/8″. To get a little more power, sometimes you can add another loop of very weak rubber.

Always remember that it is better to crash under low power than under high power, so a smaller rubber in a longer loop is your best bet for first test flights.

TYING AND LUBRICATING YOUR MOTOR

After you have cut your motor to length and tied the ends together with a square knot into which you chewed some saliva and pulled it good and tight, you might want to thread-wrap about 1/8″ after the knot and tie it tight while a friend stretches the rubber. This will help keep your knot from coming out.

Rubber lasts longer, and you can pack in more winds when you use a rubber lube on it. There are rubber lubes available from FAI, Peck, SIG, or others, but I usually prefer to make my own. You can get glycerin and also tincture of green soap from your drug store. Either mix them together 50/50 or let the green soap set in an open dish until most of the alcohol in which it is dissolved, evaporates. You can boil it out quicker, but it might catch fire when you do this. Actually, some people use just one or the

You can mix your own rubber lubricant.

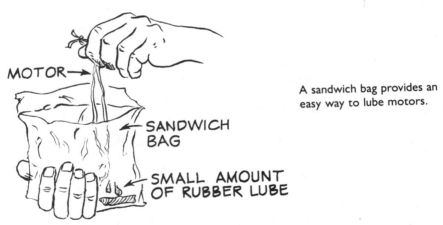

A sandwich bag provides an easy way to lube motors.

other on their rubber, and even castor oil and baby shampoo have been used in a pinch. A few modelers swear by silicone lube, but it is harder to wash off than the other products.

The neatest way to get the lube on your motor is to put a little in a resealable sandwich baggie with the motor. Work it around until it is well-lubricated. If you get it too wet, you can always blot a little off with a rag. The motor should feel slippery. If it feels dry, lube it again. You can keep extra motors in baggies (labeled as to size and length) or in old 35mm film containers with snap-on lids. Having a few extras is a good idea so that you can change easily if you get dirt on one, break one, or have the knot come undone (you can't re-tie it without washing all the lube off!).

MECHANICAL WINDERS

To wind the motor, a mechanical winder is very helpful. Peck's 5:1 winder is excellent. You can get their plastic 16:1, but I advise against it because the handles tend to break off easily. A hand drill will work, using a bent nail with the head behind the chuck jaws to prevent pulling out. However, hand drills will only wind at 3:1, which is sort of slow for this size rubber. If you use a hand drill, see how the chuck end is put together.

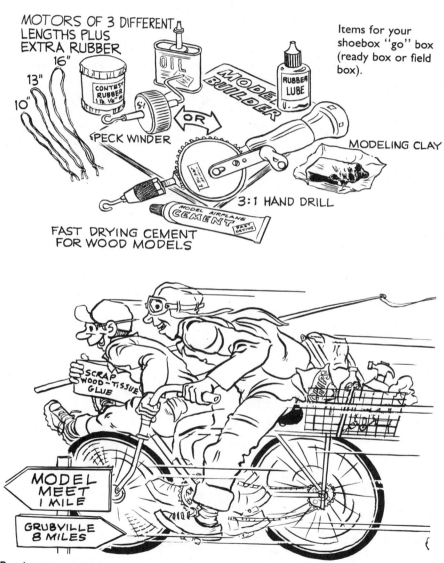

Puzzle picture: Can you figure out what they left home? Hint: It isn't tools or other support gear!

Some "cheapie" brands let the whole front end slip out when the cranked gear wheel gets loose from wear. You can get by without a winder, but it takes forever to put in over 1,000 turns.

The last stuff to throw in your "go" box of junk you are taking to the field includes a tube of quick-drying (not instant) glue, a bit of modeling clay for balance, and some extra rubber. A copy of this book will top off your equipment. Don't forget the model.

PRE-FLIGHT WORK

Check all the glued parts of the model by wiggling them a bit to see if they break off or are loose. It's easier to fix things now than out on the field.

Close one eye and hold the model out about two feet and view it under the wing. If you can see any twists, hold the wing over steam and twist in the opposite direction. Repeat this until the wings have no twists in them that you can see. Do this for each panel of the wing. Now, put a little intentional warp in the left (the one with "Sleek" printed on it) to give it a little more lift than the right wing. You are going to fly your model in left circles, so you want the left wing to drop because it is flying a little slower than the right one. Also, the model will want to roll left as

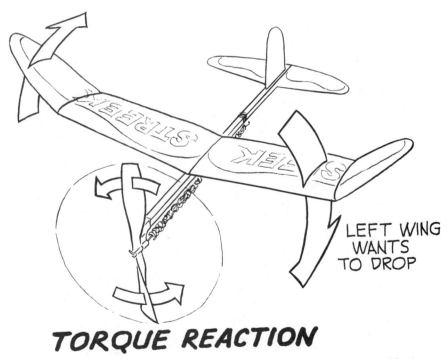

LEFT WING
WANTS
TO DROP

TORQUE REACTION

Both prop and plane are free to rotate. If given a huge prop, the plane would spin rapidly but not fly. With planes, as with people, the one hardest to move often wins!

the propeller rotates right, called *torque reaction*. The wash-in, or increase in the positive incidence, should be about .075″, the thickness of a nickel, as measured at the front of the wing about where the first "S" in "Sleek Streek" is located. While you are steaming, you might want to put just a little left in the rudder by bending the trailing edge of same to the left about .050″, the thickness of a dime. You need not do anything to the stab except make sure that it is flat.

Steam and twist just a little wash-in (about a nickel's worth) in the left wing to counter-act torque roll. Check on a flat surface as shown.

Put a drop of oil on the prop shaft where it passes through the white plastic bearing block and a drop on the place where it comes out through the front of the prop to make certain everything spins freely.

OPTIONAL TABS

On a full-size airplane, there are certain parts of the wings and tail that the pilot moves during flight to control the plane. These are called *control*

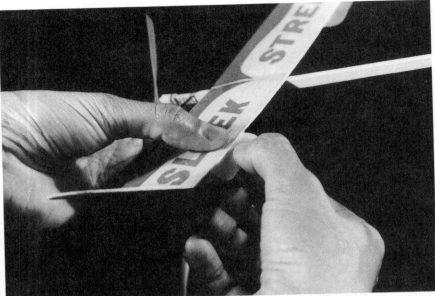

Putting in a little "down" aileron tab will give about the same effect as steaming in wash-in to correct for torque.

surfaces. They are usually parts that point down or to the side to deflect the airstream. When you put some wash-in in the wing, it was to deflect more of the air passing under the wing downward. This extra downwash resulted in forcing the wing upward a bit, giving more lift.

You could do the same thing by attaching a tab to the rear of the wing and bending it down or up to give the wing on that side of the plane more or less lift. Pioneer aviator Glenn Curtiss and others used this technique to control their roll. (The Wrights used wing-warping to do it.)

Essentially you are adding an *aileron* tab. With it, you can make flight adjustments more easily on the field. You can also add a tab on the rudder that can be bent right or left to control the *yaw* of the model, or the way the nose points. Another tab on the stab can be used as an *elevator* to con-

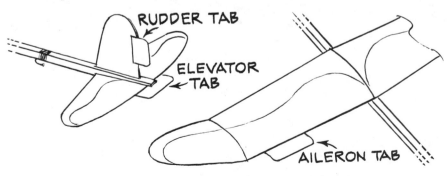

Basic control surfaces (adjustment tabs).

trol the angle of attack of the wing—giving the whole wing more or less lift.

These control tabs are options, but they will definitely make it easier for you to adjust your plane the first time out. Breathe heavily on the wing or tail part you want to bend or twist, while holding it twisted in that direction, and then let it spring back to its new position. You might have to repeat this procedure several times before it stays. Using steam or heat is better, but you probably won't have a stove on the field!

Adjustment tabs can make the model easier to get flying, but they are optional. Envelope flaps and drink-can aluminum work well.

One caution for using control tabs: Do not crease them. They are not supposed to flop up and down or be hinged like the surfaces on a full-size plane. If they do flop around, your plane will do something different each time you fly, and that is exactly what you don't want! You could even cut them out of a soda can instead of paper to make sure they don't bend too easily, but be sure to rough up the can material well so glue will stick to it. Also, you must never bend your control tabs for no reason, because they affect your flight very seriously. Never bend a tab more than about .050″ at a time; too much tab bend is worse than not enough in most cases. If, for example, you bend an aileron tab down 90° (straight down) it will give lots of drag and not much lift and will pull the wing to the rear, tightening the turn. That might be what you want in some cases, but for right now, start with all your control tabs in a neutral position—neither up, down, right, nor left. See chapter 3 for illustrations on how to adjust these tabs when the time comes.

A GOOD MODEL BOX

You should have a box to keep your model(s) in which is not only large enough to get your models in without parts hanging out, but which also

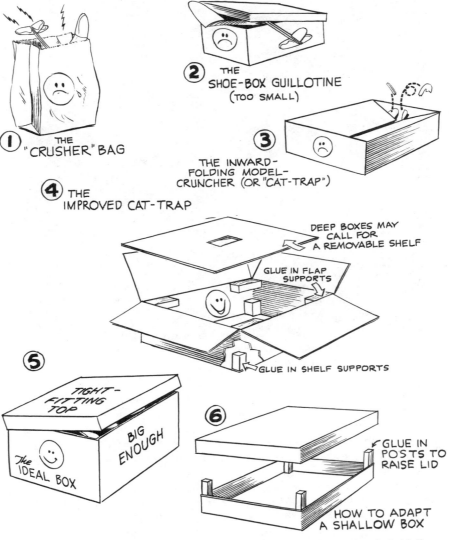

Some "don'ts" for model-protecting boxes (1-3). The right way to do it (4-6).

has a lid that can't fold down inside and injure your models. Fold-down flaps can be blocked up with a piece of wood glued inside to stop them. Never use a paper bag: It's worse than nothing and guaranteed to crush your plane. Boxes protect models from accident, cats, and rain.

Another tip: Carry your glue, winder, and other heavy stuff in your pockets or in a shoebox, but never in the same box as your models. You'll find out why the first time you drop your box and the glue goes through your tissue-covered wing!

Many a flying session has been ruined because the plane was not prepared right, protected, or because something important was left home on the workbench, so don't neglect your pre-flight work.

Two other cautions might be in order, depending on where you live. If you will be flying on a sunny day, I recommend using sunblock. When I was a kid, we used to think it was cute to get a sun tan from being out in the sun. Many of us have skin cancer today because of it. Also, if you will be chasing through tall grass or bush, protective clothing to protect from ticks which can carry Lyme disease or from rattlesnakes can save the day for you. I want you to remember that safety is no accident!

PRE-FLIGHT CHECK

 ①

ANYTHING LOOSE?
GLUE IT!

②

ANY WARPS? STEAM
'EM OUT WHILE TWISTING
IN OPPOSITE DIRECTION

④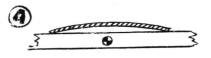

RUDDER KICKED RIGHT
OR LEFT?

WING MOUNTED LIKE THIS

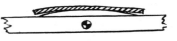

③

...OR   STAB NOT LEVEL?
FIX 'EM!

...NOT THIS

... OR THIS

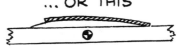

⑤

STAB MOUNTED OK?
THIS "BOOM" ANGLE OK, TOO

NOT LIKE THIS

OR THIS

⑥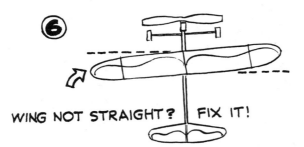

WING NOT STRAIGHT? FIX IT!

Chapter 3

Into the air

*F*or testing, it would be perfect to have a field about the size of Chicago with no trees, buildings, or wind, with about 8 inches of nice, soft grass all over it. If you are testing in a gym, it would be nice to have no roof rafters or lights hanging down; no baskets, ropes, or other junk on the sides; no drafts, with the ceiling about 300 feet high, and all of this in a round building. These places exist, but usually only in our dreams. Therefore, come as close as you can to the ideal, and let's go for it!

TIME TO FLY

A quick eyeball check over the plane to make sure no warps sneaked back in between your pre-flight check at home and the field is a good idea. If any have crept back in, breathe on the wing or tail parts heavily for about 20 seconds while twisting in the opposite direction until it stays where you want it.

Now hang your longest and weakest rubber motor on the propeller hook. If you want, use a drop of glue on the wire to close the opening, but don't glue the rubber. If you are using a winder, find the knot in the rubber and hook that end of your motor to the winder. You want the knot to be as far back on the plane as possible so it won't go "Thump! Thump! Thump!" as the motor runs down.

WINDING

If you have a winder, have your partner (whom some people call a "stooge") hold the plane by the propeller end, thumb and first finger passing over the prop and pinching the rubber on the prop hook so that it can't climb off (in case you neglected the drop of glue). Have your stooge

"Eyeball" wings for warps.

hold it so the tail is out of the way and so that the rubber, if it breaks—or you, if you get excited—will not break it accidentally. Stretch the motor out at least double its length and begin winding in a clockwise direction. If you don't know which way this is, you will soon. If nothing happens, you wound it the right way. If there is a fast "Brrrrrpppp!" sound, and your partner lets out a yip, you will know you were winding backwards and the free-wheeling ratchet device built into the front of the prop just decided to release all those backwards knots. This is not good for the prop (or your partner), so maybe you should inspect it before you try again.

For your first flight, using a $3/32''$ motor $16''$ long, you should try maybe 40 or 50 turns on your 5:1 winder, which is 200 to 250 turns in the motor. You can increase that later if the model does not go anywhere. Use less turns if you are starting with a shorter or a $1/8''$ motor, as they will be more powerful. Start walking in toward the model when you have about half of the winds packed in, arriving at it just as you put in the last turn.

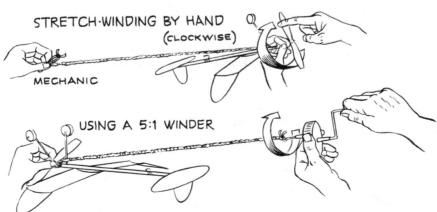

STRETCH-WINDING BY HAND
(CLOCKWISE)

MECHANIC

USING A 5:1 WINDER

Using a 5:1 winder is five times faster than winding by hand.

Grab the rubber about a half-inch from your winder's hook and back off on the winder until you have a nice loop to hook over the model's rear motor hook. This is where the knot should be. **Note:** Be especially careful to stay away from the tail while you are hooking on the rubber! It is easy to get so occupied with doing one thing that you bump your tailfeathers. They will either break off or re-arrange your adjustments for you. Have your mechanic (sounds better than "stooge," doesn't it?) put the winder back in the box immediately. If you don't step on it or lose it, it will be a great help on your next flight.

READY? EASY DOES IT . . .

If you don't have a winder, you can still wind up long motors; it just takes a little longer. Give the tail-end of the rubber motor, with the knot, to your partner or loop it over something solid to hold it. At the same time, hold your model by the white nose bearing, packing in the turns one at a time by turning the prop with your finger.

When a scientist performs an experiment, he/she has to be a good observer! You cannot figure out what happened unless you saw what happened and remember it long enough to do something about it. I've asked kids what happened when they come up with an airplane that won't fly and they've said, ". . . it went up and down." Oh really? The picture that comes to mind is that of a yo-yo. If the flier said, "The nose went up and then fell towards the ground," I might have diagnosed a stall, perhaps caused by too hard a launch, a heavy tail, or too much "up" elevator. Whether your model veered off to the left or to the right makes a world of difference as to what you do to correct it! Check the troubleshooting chart in this chapter to see if you can find exactly what your model did, and then make a correction. It is a good idea to make only one adjustment at a time so you will know what made the difference.

Never wind the model up fully until it is flying nicely. A model that crashes at high speed with a fully-wound motor will often become very, very short. Some have been known to return themselves to kit form. One thing to remember is that your model will probably not fly well on the first few flights. It's the little adjustments and changes you make intelligently, called *trimming out* that will make it fly.

WHAT DOES A GOOD FLIGHT LOOK LIKE?

I like to see a model fly in left-hand (counterclockwise) circles about 20 paces across, while spiraling upward under power, followed by a gentle power-off glide in left-hand circles, also returning to earth with no stalls or dives. This is called a "left-left" pattern.

Left is the normal way the model wants to roll because left is opposite the prop rotation direction. It kills a little of the lift when the model is rotating hard left under the beginning-of-the-flight power burst. You can always take the prop shaft part of the plastic nose bearing and tweak it a bit to the right if you want a wider left-hand circle and more climb. The left rudder adjustment (about .050″) makes it want to go left, while the wash-in of about .075″ in the LH wing panel keeps the left wing up in the turn. The nose block has a bit of right thrust built in when you get it. On indoor models, launch on the side of the building, allowing the model to go into its left circle without hitting a wall.

Hand-launching technique: Hold model as shown, let prop run one second, then launch gently with the nose raised slightly above the horizon. Do not throw it, and do not launch it straight up.

For a hand launch, hold the model as shown in the photograph. Let the prop run one second, and launch it gently, with the nose raised slightly above the horizon. **Note:** Do not throw the model, and do not launch it straight up.

On a gym floor or on a hard surface outdoors, you can R.O.G. if you built the version with landing gear. To do this, hold the model with the thumb and first finger just behind the wing from the top. Let the prop start, and then let go of the model. If there is a wind blowing, it is a good idea to aim the model not directly into the wind, but a little to the right, and be sure and let the prop run just a little longer before you let go of the

Kris Samonas demonstrates the correct way to R.O.G. a model. Aim a little to the right of the oncoming wind, let the prop start spinning for one second, and then release the plane.

model. The reason for this is the torque of the twisting effect of the prop at the beginning of the prop run is so strong that is might turn your model too far left, letting the wind get under your right wing-tip and turning your plane upside down. A good R.O.G. will look very realistic, with the model going into a nice, smooth, left-hand spiral upstairs to cuddle the cumulus.

Note the follow-through—getting hands away from the model. The model should pick up ground speed and then lift off.

As the model get airborne, it banks left into its climb with the left wing low due to torque from the fully wound motor.

FIXING THE PROBLEMS WITH TAB ADJUSTMENTS

Remember: Being a good observer is the most important thing there is when it comes to making a plane fly well. First, recognize a *dive* for what it is. Throwing a rock is a good example: It starts down as soon as it leaves your hand. A *stall* is made up of three parts:

1. The model climbs a bit too steeply or zooms up.

2. It slows down a bit as the air breaks away from the top of the wing, due to its too-steep angle of attack.

3. The nose falls toward the ground, a *dive*.

As the model drops earthward, the angle of attack decreases, the model starts flying again, and then repeats the three steps again, sort of like a roller coaster.

A *spiral dive* happens when the plane starts banking (rolling to the side by dropping one wing tip or the other) and keeps turning toward the ground until it crashes. You have to note whether it is a spiral dive to the right or the left in order to make the adjustments. **Note:** when we say "right" or "left," we are pretending there is a pilot in the plane, and it is to the pilot's right or left.

Add wash-in to the left wing if your model has a tendency to roll too much to the left.

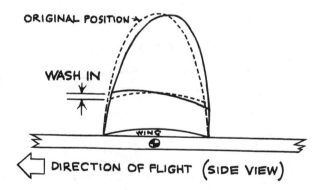

Add wash-out to the side of the wing toward which you want the model to roll.

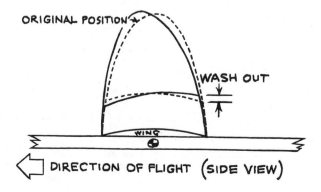

Be sure to take a look at the illustrations and troubleshooting chart before you make adjustments.

LE/TE adjustments

If your model has a tendency to roll too much to the left on launch—when the motor is delivering very high power—add some wash-in to the left wing by twisting its LE up. On the other hand, washing out either side of the wing will push that side down and help the plane roll toward that side, or stop that side from lifting so much.

Aileron tabs

If you have installed aileron tabs, bending one up forces that wing down, and the model will roll toward that side. Bending an aileron tab down raises the wing on that side, giving the same effect as wash-in (p. 35).

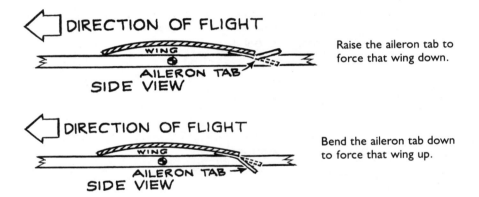

Raise the aileron tab to force that wing down.

Bend the aileron tab down to force that wing up.

Elevator tabs

Bending an elevator tab up shoves the tail down, which increases the angle of attack of the wing, gives more lift, and cures a dive. In turn, bending an elevator tab down shoves the tail of the model up, which decreases the angle of attack, reduces lift, and reduces stalling or looping tendencies in the model.

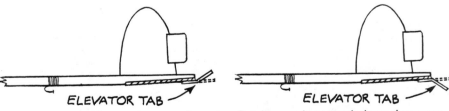

Bend an elevator tab up to increase the angle of attack of the wing and to cure a dive.

Bending an elevator tab down decreases the angle of attack and reduces stalling or looping.

Rudder tabs

Bending the rudder to the left forces the tail to go right and the nose to go left. Bending the rudder to the right shoves the tail left—which, of course, makes the nose go to the right.

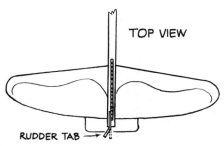

Bending the rudder to the left forces the tail to go right and the nose to go left.

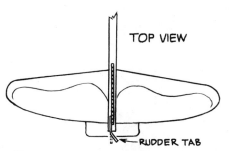

Bending the rudder to the right shoves the tail left and the nose right.

HOW FAR IS A LITTLE?

Before you actually bend those tabs, you should know how far to go. A little for one person is a lot for somebody else. Some common items you might have on hand will give you a better idea of how far an adjustment needs to be bent. A credit card is about .030″ (thirty thousandths of an inch) thick. This is my idea of a "little." A dime is about .050″ thick (a little more), and a nickel is about .075″ thick (a lot). Of course you'll be guessing, but at least you'll be in the ballpark, instead of just trying to read my mind. Take the troubleshooting chart with you to the flying field to help you over the rough spots.

If you haven't added the aileron, elevator, and rudder trim tabs, you'll be breathing on balsa and twisting a lot to deflect the air in the right directions. You'll have to check those balsa twists after every flight, though, as balsa has a habit of going back to its original position. Steaming is more permanent.

TROUBLESHOOTING CHART
No wind blowing; model launch normal

The Problem	What Might Fix It
NORMAL CLIMB / NORMAL GLIDE / DIVE Model dives straight in.	1. Bend the trailing edge of the stab or the elevator tab up 0.30 inches. 2. Add a bit of modeling clay about the size of half a pea to the tail. 3. As a last resort, reglue the wing 1/2" farther forward.
"ROLLER-COASTER" STALLS (PORPOISING) / SEVERE STALL Model stalls (Nose first goes up, hesitates slightly, then drops to a dive; roller-coaster).	1. Bend the trailing edge of the stab or the elevator down .030 inches. 2. If the model wasn't turning, bend the rear of the rudder or the rudder tab about .030 inches left (as seen from the rear). 3. Try a bit of modeling clay about the size of a pea on the nose, as far forward as it will go.
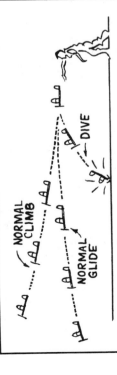 Model refuses to fly left, even though you try everything.	1. Go with the flow—fly right. Why fight it? You might have built it as a RH model without knowing it.

1. Hold the model at arm's length. Close one eye and see if the wings are warped. The right wing should be untwisted, but the left should have about .070 inches wash-in. If too much wash-in, breathe on it and twist in opposite direction. Recheck.

2. Bend rear of rudder or tab about .030 inches to the left.

3. Bend the trailing edge of the stab up about .050 inches.

4. Add about a half a pea of clay to the tail.

5. Bend right aileron tab down .050 inches and left tab up .070 inches.

1. Remember that the rubber spinning the prop makes the plane roll left. When the motor runs down, this force is missing. Try adjusting the model so that it glides well, and then play with the prop-shaft part of the nose bearing. Twisting it a little right will open up a too-tight left turn; a little left will turn a straight climb into a left circle, etc.

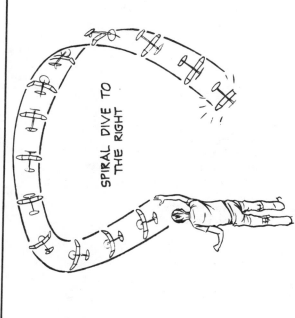

Spiral dive to the right: Model raises its left wing and finally crashes to the right.

The model flies great (power phases) until it runs out of power, then it dives, stalls, or goes straight.

CHANGING THE CG

If you built your model and balanced it correctly before adding the wing, no changes in the center of gravity should be necessary. However, sometimes differences in motor weights when you add a longer motor might require a little nose weight to rebalance the model. Sometimes a model

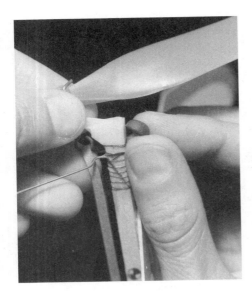

Adding a little modeling clay for nose ballast can correct a tendency to stall by moving the CG forward.

Adding a little modeling clay to the tail can move the CG aft, curing a tendency to dive. Careful! A tiny bit goes a long way, due to the long "arm" leverage of the tail on the Sleek Streek.

that refuses to climb can be coaxed into doing so by making the tail just a wee bit heavier. Modeling clay is ideal for adding weight, and it does not take much. Never add weight to both the nose and tail; that does nothing to adjust the balance, but just makes the model heavier. Sometimes, adding just a little clay to one wing tip or the other may help a model turn, but this should only be tried after all other methods have failed. Again, adding weight to both wingtips just cancels out the effect you are trying to achieve.

SOME FINAL NOTES

Try not to re-fly bad flights. Do something differently on the flight following a bad flight. Oil the propeller bearing every few flights to keep dry plastic from rubbing on dry plastic. The rubber motor gets "tired" through being wound fully, and it should be changed now and then. If it's nicked or torn, it should be thrown out. However, a motor that has gotten stretched will recuperate with a half-hour's rest.

Keep rubber in a container that keeps out air and light, both of which ruin rubber. Re-lube the rubber motor when it feels dry. You can run a dry motor, but you will not be able to pack in as many winds, and the rubber will tear easily.

The only advantage of a dry motor is that sand does not stick to it as easily! If you get grit all over your motor, change it or expect it to break soon. It can be washed off and re-lubed if you have water handy. If you get dirt into the nose bearing and the prop feels funny when you turn it, better wash that off with water without getting any on the wings and tail. Water is the enemy of sheet balsa; it gives you warps, unless you pin the wet part to a flat board and let it dry for a day or two.

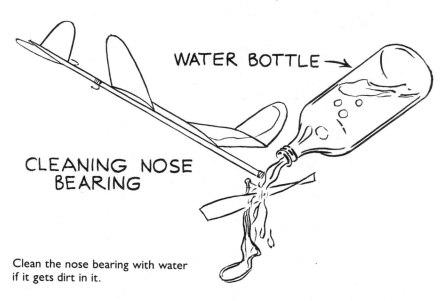

WATER BOTTLE →

CLEANING NOSE BEARING

Clean the nose bearing with water if it gets dirt in it.

I suggest that you don't risk your life for your model, as much as you love it. Running in front of cars when your plane is flying over a street is a maneuver that can land you in the hospital, and climbing trees can have the same result. Throwing things at it to dislodge it is OK, as long as no one is standing underneath to get beaned by what you threw up at it. Some guys use a fishline with a weight on the end to throw over a branch and then yanking on the line to shake the model out.

Some of you might wonder why you should make your model fly in circles, but after you see it fly, you'll know why. If it went straight, it would leave most flying fields and certainly would not do well inside a gym! Also, by circling, you can keep it inside a *thermal*, or rising bubble of warm air caused by the sun heating the ground. Birds in thermals do

Use a weighted line to hook and shake a branch (to retrieve your model from a tree).

not have to move their wings to stay up or to climb because the air around them is going up faster than they are gliding down. Your model can do the same thing. Losing your first model up and out of sight can be a real thrill!

INDOOR FLYING

If you will be flying in a gym, you might want to spend some time with sandpaper making your model as light as possible, or even making a new one out of lighter balsa from your hobby dealer (SIG Very Light Contest Balsa is a good choice). If your model is light enough, you will be able to use lighter rubber, such as $1/16''$, instead of $3/32''$. If you use heavier rubber, you can cut its power down to get long runs by making longer loops, by twisting the prop blades into higher pitch (so you will see more blade when looking at it from the side—more air resistance will make it turn slower). Twisting the blades to low pitch (less air resistance as seen from the side) will make the prop spin faster and give more power, but with a shorter run. This is useful if your $1/16''$ motor is not quite powerful enough to get you to the ceiling. I have had over two-minute flights in regular gyms with Sleek Streeks by experimenting with different motors and prop pitches.

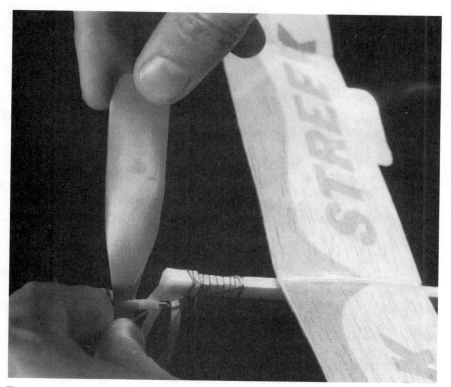

Twisting more pitch into a plastic prop slows it down and makes the motor run longer with less power. It also helps keep the model from hitting the ceiling indoors.

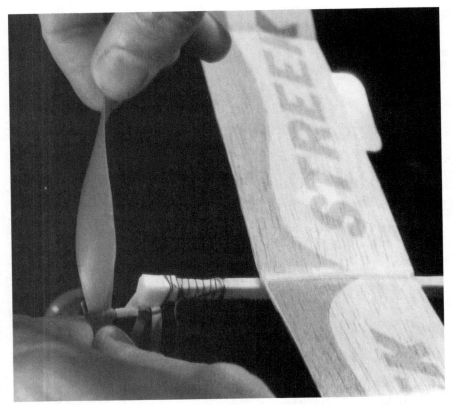

Twisting less pitch into the prop lets it run faster, letting the motor get more power but for a shorter time. This is a good adjustment to gain altitude with a weak motor.

Outdoors, your flight possibilities are unlimited. Walt Mooney once saw a Sleek Streek fly by a glider he was piloting at 7,000 feet! Some people have been known to put their phone number on their models when they start flying that well.

REPAIRS

I hope you don't need any, but it's a good idea to have your tube of cellulose glue (Testors "green tube" fast-drying) with you. If your wing or tail parts are cracked, **do not** smear glue on the top or bottom of the wing. It will warp your model when it dries and shrinks up. Instead, break off the cracked portion, spread cement on the edge of where it broke. Do the same for the place it broke away from, wiping off the excess after you put it back together. The only glue that remains should be what is in the joint.

If the fuselage snaps, generally just gluing the parts back together is not enough. Gluing a sliver of balsa or a toothpick on each side of the fuselage after it is glued back and wrapping the whole thing with thread (winds spaced slightly apart and then glue rubbed into it to hold it in

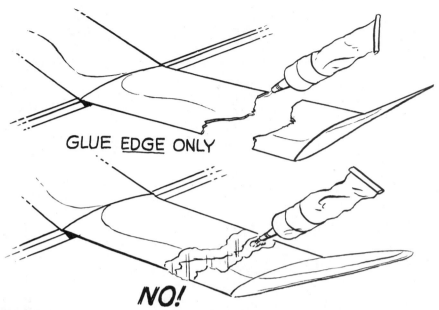

GLUE EDGE ONLY

NO!

Don't smear cement around on thin balsa flying surfaces. Cellulose glue shrinks as it dries and will warp your model like a Pringle. Double-glue the contacting edges of broken parts only.

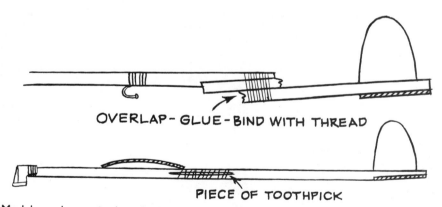

OVERLAP-GLUE-BIND WITH THREAD

PIECE OF TOOTHPICK

Models can be repaired on the field as shown. "Instant" glues are not recommended, for safety reasons.

place) will help, but this also adds weight. Sometimes you can just make an overlap joint by putting one part over the other for a little ways with glue in between and then doing the thread-wrap/rub-in-glue routine. Check and make sure it's straight before it dries (15–20 minutes). Crude, but it might last you until you get home where you can fix it with a new part.

TWO VIEWS ON INSTANT GLUE

Most of you are familiar with cyanoacrylate glue, often called "instant glue," "Super Glue," or "Crazy Glue." My recommendation is that you stay as far away from this stuff as possible. Most kids I know get careless now and then, as do adults. Three adult friends, all expert modelers, have had to go to the hospital emergency room with this glue getting into their eyes. Yes, I know that YOU are smarter than they were, and that it would never happen to YOU, but the warning stands anyway.

My main concern is with your having fun, and the convenience of having a glue that "dries" in seconds is not worth the chance you are taking. There are all kinds of ways it can get in your eye: having some on a finger and rubbing your eye, having a bottle on the edge of a table and bending down to pick up something from the floor, knocking it over and having a drop fly through the air, having some pop out from between two surfaces when it "kicks". . . There are probably other ways, but I have seen these happen. If you absolutely cannot live without using it, please wear safety glasses and use the thicker types which are not quite as easy to err with. In this book, Testors "green tube" fast-drying glue is recommended for both building and repairing.

The case *for* cyanoacrylate glue is that it sets very quickly, penetrates well, and forms a joint many times stronger than the original wood. Generally a *kicker* is sprayed on it to speed the setting and this should be treated as dangerous around eyes. Editor Bill Northrop of *Model Builder* magazine feels that this type of glue is no more dangerous than high-revving engines turning sharp propellers, sharp model knives, and razor blades—or even "sniffing" model cement or dope. His position is that we need to recognize the hazard and avoid it, to take the appropriate precautions. The final decision is up to you.

Chapter 4

Building the
Peck R.O.G.

*T*here's just something about opening a model kit that makes you want to start sticking balsa together! Most kids make the mistake of starting with something way too hard, usually something from World War II that had a pretty picture on the kit box cover right over the words "Flying Model." This, of course, is sheer nonsense in most cases, as most of them get so messed up by the time they are finished (if they ever are) that you couldn't tell if it was a B-24 or a Piper Cub. Fly? With the wood and heavy plastic formed parts most kits contain, even an expert could not get it to fly, let alone a beginner. Well, you're not a beginner anymore if you've successfully made and flown the *Sleek Streek.* You now know some things to do and some things not to do. Time to go on to something harder that will get us into stick-and-tissue building with a proven flier. The plane is the *Peck R.O.G.*, and its flight will match those of the *Sleek Streek* hop-up (which you have probably lost in a thermal or in a tree by now).

It might seem like I'm taking a long time just to help you build a couple of simple models, but just remember: The basics you are learning now apply to all models that you might build, and even to full-size airplanes. Remember how you balanced your Sleek Streek R.O.G. so the center of lift and the center of gravity would be close to each other? Well, they have to balance full-size airplanes, too!

THE BUILDING BOARD

One of the things that makes life easier for a model builder is the building board—a flat board on which the pieces of the model are pinned down to dry, ensuring an untwisted framework. It should be soft enough to stick pins into easily. "Celotex" wallboard, 1/2" thick, is the best thing I have

The Peck R.O.G. is an excellent introduction to stick-and-tissue modeling, and it is a good flier to boot.

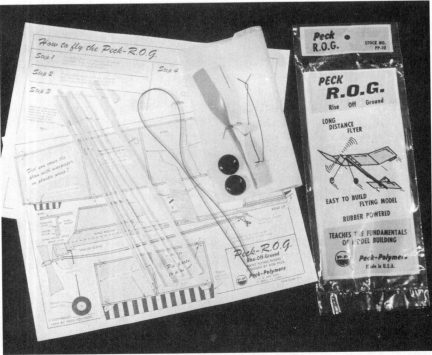

Check through the kit to see if all your parts are there (don't lose any!), and read all plans and instructions before starting to build.

found. It's made of soft fibers and painted white on one side. You can get it at your local lumber yard in a sheet four feet wide by eight feet long for not much more than you'll pay a hobby dealer for a small piece—about $8 or so.

You can cut up the wallboard with a knife to get it into the family car. Just make a cut part way through and you can break it the rest of the way. You can sell the extra to friends for their models or, as I do, have three or four models going at the same time. You'd be amazed how many 14"×24" boards will partly-built models on them you can stuff under the bed! Or you can pin pictures on it and use it for a dart board.

If you can't get Celotex, heavy cardboard like they use to pack refrigerators in is the next best thing. If it isn't flat, maybe you'd like to glue it down to a board or old bench that is. Flat is where it's at.

THE CUTTING BOARD

There are two reasons for not cutting on your Mom's formica kitchen table. One of them is that you'll dull your knife, and I'll let you guess what the other one is. One modeler I know uses old phone books to cut parts out on, tearing off a few pages when the surface gets too cut up.

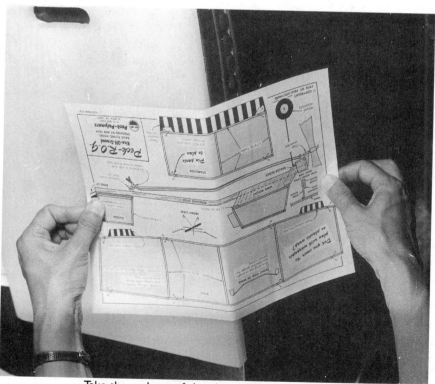

Take the curl out of the plan on the edge of a table.

Another friend uses old linoleum tiles. I like heavy, solid cardboard about 1/8″ thick, but I have no idea where you can get it outside of art supply stores. Don't cut on your plan or building board if you can help it, because it really makes a mess. Celotex dulls your knife or razor blade, too.

THE PLAN

Model airplanes are built right on top of the plans. That is because many structures like wings and tail parts have to be pretty flat to fly. If your plan has been rolled or folded up, it may be hard to use, so hold it by the ends and run it over the edge of a table a few times, putting a little pressure on it to "erase" the curl and ridges. You can iron it if you aren't in a hurry! Then, tape it down on your building board, pulling out the wrinkles. I usually use frosted Magic Mending tape, but for the pictures I used vinyl black electrical tape so you could see it. Pinning your plan down isn't good enough, because it won't stay flat.

Once you have taped the plan down to the building board, tape a layer of plastic wrap over the plan to keep the parts from getting glued to the plan. I used to use waxed paper, but the wax seems to get into the glued joints and keeps the glue from drying as well as it should. Some older modelers rub the parts of the plan that will come in contact with glue with a wax candle or a dry bar of soap. I even knew a guy once who

If you forget to use plastic wrap over the plan, your parts will get stuck to the paper (and that can be very embarrassing!).

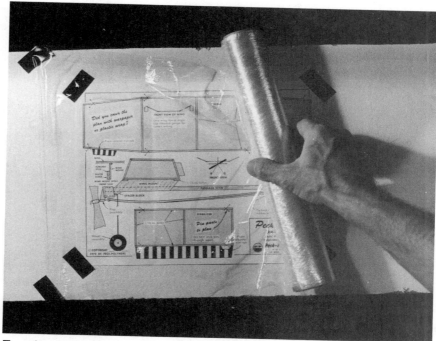

Tape the plastic wrap over the plan. (Black tape used here for illustration purposes only.)

glued all his parts to the plan on purpose so he would have less covering to do. He just cut out around the edges of the wing and tail parts! You can do this if you don't intend to fly the model, but you really don't need all that weight on a Peck R.O.G.: It will weigh too much to fly for long. Some kids like to try to use the plastic wrap to cover the model, but glue doesn't want to stick to it. After all, that's why you put it over the plan in the first place, right?

PICKING THE STICKS

Yesterday, I sorted through about 2,000 1/16"-square balsa sticks that were similar to those you'll find in your Peck R.O.G. kit. I broke about a hundred of them in doing so, since I test them by bending them and "feeling" their strength. I try to keep the kits I make up for my aviation classes at school free from weak balsa. Beginners often use strong sticks where weak ones would do and put in weak ones where they need strength! My students do the final sorting with what I call a "stick sorter"—hanging a weight, in this case a dime, on the end of the stick and seeing how far it bends down. The ones that bend the least are the ones they'll use for the leading edges and trailing edges. The lighter ones are used as *ribs* (the parts that go between the LE and TE) and as tail parts where we try to save weight. Even on the tail assembly, a bit more stiffness on the LE and TE

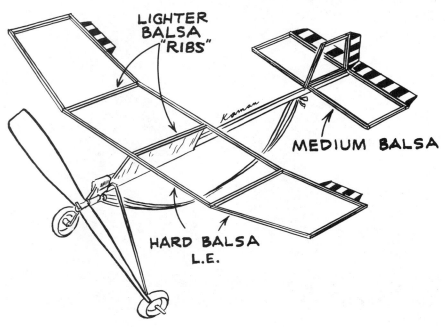

LIGHTER BALSA "RIBS"

MEDIUM BALSA

HARD BALSA L.E.

It's wise to use the harder, heavier balsa where the model takes its knocks. Lighter balsa, where you can get away with it, saves weight.

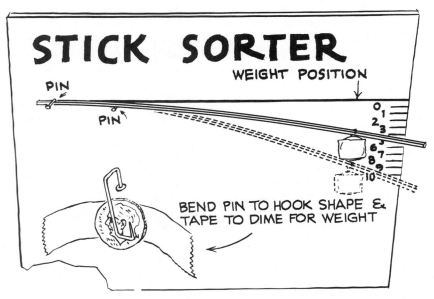

STICK SORTER

WEIGHT POSITION

PIN

PIN

0
1
2
3
5
6
7
8
9
10

BEND PIN TO HOOK SHAPE & TAPE TO DIME FOR WEIGHT

A simple stick sorter can help you separate the heavy from the light sticks.

helps keep it flat. This will become much more important when we get into doing more advanced models later.

You can make your own sorter out of a piece of cardboard with something on it (two pins, a bit of soda straw, a clothespin, etc.) to hold one end of a stick. The end of the stick can be weighted with a small nut or a dime hung on with a tape-on pin or wire bent into a hook. The little lines are just for comparison to see which ones bend down the farthest.

You might think that the weight of the sticks is no big deal, but I have seen three out of four kids use the spruce (harder wood) tail boom included in a popular beginner's kit for a wing part, and use the soft balsa for the tail boom, which takes a lot of stress in a crash. Guess where these models always break? Now is the time to start thinking about what takes the most abuse, and where the stiffer, stronger wood needs to be. Once you start doing this, you might find that some of the wood in the kit is no good at all, and you might have to start keeping some stock of your own for these emergencies. Always save your extra wood!

RAZOR BLADES OR MODEL KNIVES?

When I was a kid, the only choice we had for cutting balsa was to use a razor blade, usually a dull one Dad had used shaving. They made them out of real carbon steel in those days and they cut pretty well, even after being used for their intended purpose. Today's blades that you get in the market are garbage, having softer, but nonrusting metal mixed in with the steel, making them better for shaving, but lousy for anything else. You can get "industrial" single-edge blades (which are still pretty good) at paint or hardware stores. However, if it says "stainless" on the package, pass it by.

I like razor blades better than model knives for cutting balsa sticks off for two reasons. First, razor blades are cheap and can be thrown away when they are dull. Most modelers use the "throw-away" model knife blades far too long, and almost no one knows how to sharpen one anymore. The second reason is that a razor blade is easier to line up for a straight cut than some knife you hold like a pencil, off at some funny angle. By sighting straight down over the blade, you can make those cuts

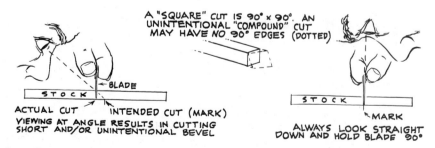

By sighting straight down over the blade, you can make cuts right where you want them to be and perfectly square.

square and right where you want them. Model knives comes in handy when you need to cut out ribs and other parts from sheet balsa on more complicated models.

If you tend to be accident-prone, I suggest that you wrap masking tape around the half of the razor blade you are not cutting with, and never use double-edge blades.

Remember, you don't want to actually cut through the plans on your building board. Instead, mark the point where you want to cut with the razor blade while the piece is on the plan. Then move the piece to the cutting board to actually cut through it.

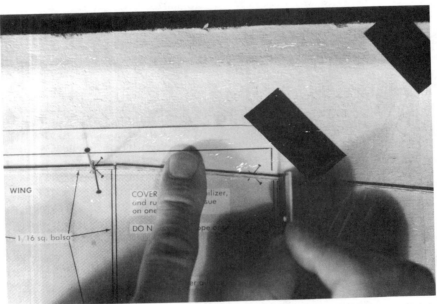

Mark the parts the length you want to cut them right over the plan. Be sure not to cut through the plan—or yourself!

BUILDING THE WING

You should build the wing right over the outline shown on the plan. Even though the wing tips are raised to give dihedral on the finished wing, build the whole thing flat for now. The LE is made in three pieces. Cut them from the hardest sticks you have in your kit, and pin them down to the plastic-wrap-covered plan. Do not stick pins through the wood!

Use the "X-ing" method (shown in the illustration) of leaning the pins onto the wood in pairs, applying just enough pressure to hold the piece down without denting the wood and weakening it. You can also use T-pins (dressmaker's pins) for this. Don't glue the leading edge parts together now, or you'll just have to cut them apart later when it's dihedral time.

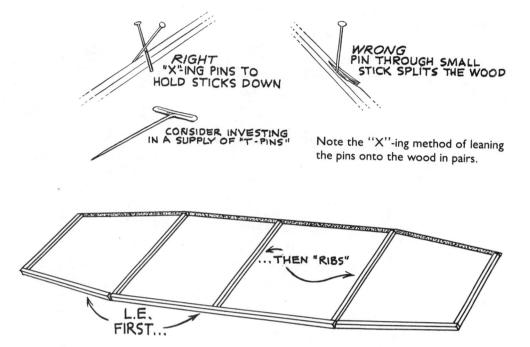

RIGHT
"X"-ING PINS TO
HOLD STICKS DOWN

WRONG
PIN THROUGH SMALL
STICK SPLITS THE WOOD

CONSIDER INVESTING
IN A SUPPLY OF "T-PINS"

Note the "X"-ing method of leaning the pins onto the wood in pairs.

...THEN "RIBS"

L.E.
FIRST...

Start the wing by placing the LE and then putting the ribs in. The trailing edge goes on last.

Now cut all the *ribs* (sticks that connect the LE to the TE) from the softer sticks in the kit: The ribs don't take as much stress as the LE. Glue each one to the LE and pin it down, leaving the TE to be added later. Now trim any of the ribs that seem to be a bit too long so they are all the same length. The TE should touch each one of them without having to be bent in or have a gap filled. You can never stretch short sticks or ribs, so my suggestion to you is to always cut them a tiny bit long and trim them for a perfect, installed fit. If you force a part in, it may wind up giving you a twisted structure later on when you take the wing off the board.

It's also a bad idea to try and fill up the space between two poorly-fitted parts with glue. Glue adds weight, it takes forever to dry, it looks bad, etc. Either make another part the right length, or glue in a bit of scrap balsa to fill in the gap.

If you have several parts to cut to the same length, you can cut one the right size and then use it as a pattern to cut the others. If you do this, be sure and use the *same* part as your pattern each time and not the one you just cut off with it. This prevents the length "growing" as you go along.

Again, remember to cut on your cutting board and not on the plan. You might want to use the plan again.

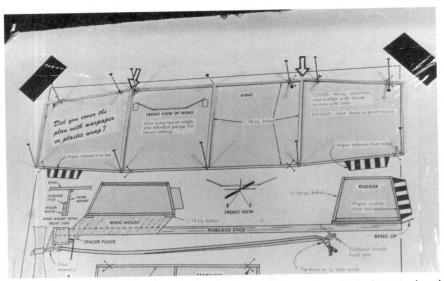

The ribs are held against the LE for the time being while the glue dries. A pin is placed where the TE will go later.

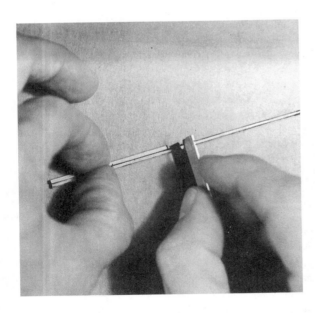

Cutting several parts all the same length is easy if you use the first one as a pattern. Always use the same part to measure the others.

Another tip for good cutting is to keep the blade perfectly in line with the line on the plan, both side-to-side and up-and down. Pushing the razor blade slightly forward as you cut generally works better than just crushing straight down. Remember, a sharp blade is essential. Once you get glue on the edge or try to cut paper clips, toss out the blade and get a new one.

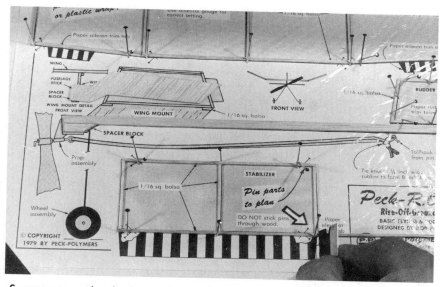

Some parts can be glued on and trimmed to length later, when the structure is dry.

CUT WHERE THE PLAN SHOWS

"Jeez! I've cut it off three times and it's *still* too short!" Hard as it might be for you to believe, some people are almost that dumb when it comes to cutting parts off right. Study the plan carefully and try to figure out why a part should be cut where it is. I once had a kid cut a leading edge off every time it came to a rib. When I asked him why, he just said, "I dunno."

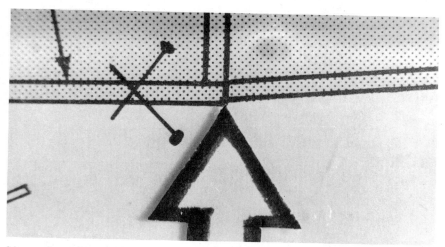

Note where the parts meet (line straight across stick). If there is no line, don't cut. The arrow here points to the dihedral break in the LE, which doesn't get glued on until wing tips are raised later.

Brain damage? Not really. He just didn't stop to think that the purpose of the LE is to add strength to the front of the wing, and that cutting it up made the LE weak. When the guy who drew the plan doesn't show a line crossing the stick, don't cut it off! Often when two sticks cross, you will have to decide which gets cut and which goes on through the intersection. Careful inspection of the plan will generally tell you which one the designer of the plan wanted left in one piece. Trust that designer. Ninety-nine times out of a hundred, he or she had a reason for drawing it that way, and it might be just a little better than leaving it to chance.

GLUING TECHNIQUE

You may use white glue such as Elmer's, or an aliphatic resin glue such as Titebond, or the yellowish "carpenter's wood glue." If you use Testor's "green tube" cellulose cement (Fast-Drying Cement for Wood Models) as I do, you will have to work faster.

Put a drop on the end of each rib and then dab the drop quickly against the place on the LE where it goes. (Getting glue on both halves of the joint is important. If you wait too long, it will be too dry to stick well to the LE.) Then, put the piece in place and "X" a couple of pins to hold it, with a third pin pushing it toward the glue joint at the place where the

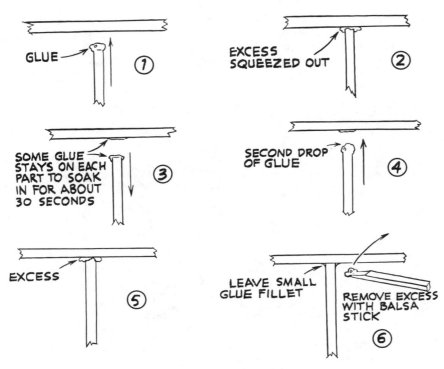

How to double-glue a joint.

TE will go later. With the end of a stick, wipe up any extra glue that squeezes out of the joint to save weight and make it dry faster.

Blobs of glue get a skin on them like an egg and do not dry inside. If you use white glue, the easy way is to put a few drops on the plastic wrap and dip the ends of the sticks in it. Allow the wing to dry "as-is" for about a half hour or so while you go on and do the stabilizer and rudder.

Pay attention to which sticks run all the way to the edge and which butt into the part they are glued to. Study the plan carefully instead of just assuming you know. Sometimes it makes a difference in strength.

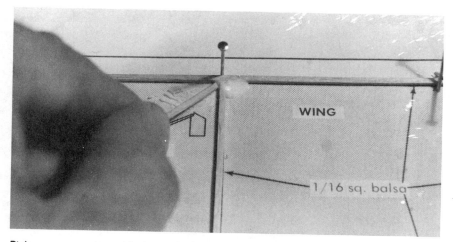

Pick up excess glue with the end of a stick or it will add unnecessary weight to the model.

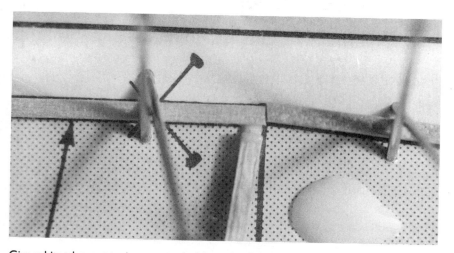

Give white glue extra time to work. Note the drop of glue on the plastic wrap, used to dip the ends of the ribs in. Double-gluing always helps.

Notice on the plan, for example, the way the rudder is drawn. Do you see that the LE and TE of the rudder go all the way to the part with the little angled lines on it (the side view of where the stab will go), and the top and bottom pieces of the rudder glue up against them. This will help keep the rudder from tipping to the side and breaking off.

Before you glue the TE on the stab, go back and put the TE on the wing, as the ribs are probably dry.

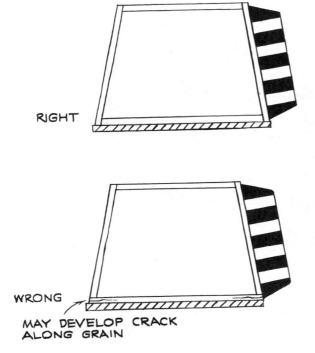

RIGHT

WRONG

MAY DEVELOP CRACK ALONG GRAIN

Plans are usually drawn the way they are for a reason. Notice the difference between the way the designer drew the line (top) and the way someone made it wrong (bottom).

FINISHING THE FRAMEWORK

Take out the pins that you used to apply pressure to the ends of the wing ribs while they were drying. Cut the TE sections, and glue them to the rear ends of the ribs, using the end of a piece of scrap stick to apply a drop of glue to each joint before you touch the balsa together. Don't glue the TE sections to each other (remember how you did the LE?). A pin behind the TE at each rib location will hold enough pressure against it to keep it in place while it dries. Do the same thing with the stab.

Fuselage

While the wing and tail frames are drying for about a half hour, study the side view of the fuselage assembly on the plan. Glue the spacer block to the front end of the fuselage stick—on the narrow edge, not the fat part.

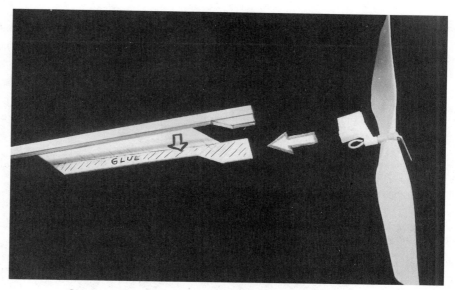

Glue long fuselage stick and spacer-block to the pylon as shown.

Hold it over the plan to see if you did it right. Then take the 1/16"-thick balsa-sheet wing mount and glue it on the left (pilot's left) side of the fuselage. Do you notice the broken or dotted line on the plan that runs just under the words "wing mount"? That means that you cannot see that part of the fuselage when the wing mount is in place, and that's how you know it is on the left side.

Now glue the two tiny bits of 1/16" square balsa on to the wing mount as shown in the plan. As you can see, these are on the left, too; they have no dotted lines. Now look at the center drawing on your plan by the left hand margin entitled "Wing Mount Detail, Front View." What you are

Adding a short length of 1/16" square stick on the upper part of the pylon as shown gives the wing more to glue to. Do both sides and each end.

looking at is what you just did, as seen from the front (minus the wing, of course). If your work doesn't look like the plan, try to figure out what went wrong. Set the parts aside to dry.

Dihedral

Making sure the wing tip (three sticks glued together) is dry, slide your razor blade or a table knife blade between the sticks and the plastic wrap. Pry the parts off gently. They do stick just a little, and you could break the tiny structure easily.

Now glue the tip on at an angle, with the tips raised one inch higher than the center section of the wing. To make the LE and TE fit better, you will need to sand just a little angle where they join.

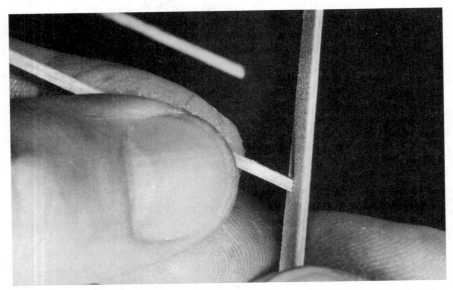

Sand a small bevel on each LE part of the wing tip section to match the dihedral angle. Grasp the stick close to where you are sanding or it will break.

DIHEDRAL JOINT

WRONG
(NOTE GAP)

RIGHT
(BEVEL)

Glue the tip on at an angle, with the tips raised one inch higher than the center section of the wing.

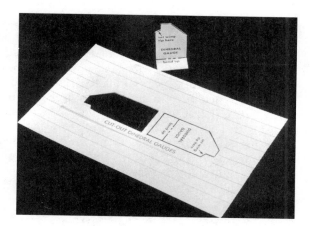

Glue the dihedral gauges from the plan to a piece of card stock or scrap balsa and cut them out.

To prop the tips up while drying, glue the dihedral gauges (on the instruction sheet) to thin card stock and cut them out, using both of them on the same tip a little apart to keep the tip level. You could also use a one-inch block (I use a VHS video cassette under each tip) if you are lazy. If you use Testors on the dihedral joints, allow it to dry at least an hour. White glue takes overnight.

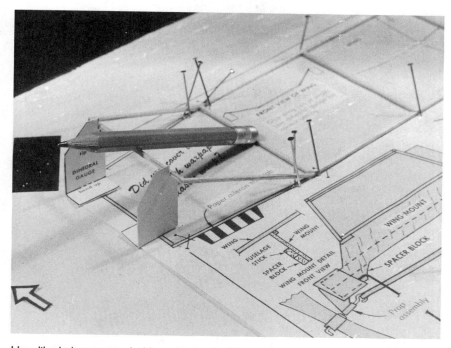

Use dihedral gauges to hold up tip panel while gluing; using two on one tip ensures a nontwisted structure. Let dry at least an hour.

Front end

Slip the wheels on the wire landing gear wire and bend up the ends with pliers to hold them on. Needlenose pliers work best. I had to cut about 1/8″ off each end of my wire before I bent it up, but yours might be just right. Slip the landing gear wire into the groove in the white nose bearing, and try the fit onto the end of the fuselage where the spacer block is. Chances are that it will not fit, so take your sanding block or an emery board and sand just a little at a time off the spacer block until the nose block just shoves on snugly. Don't sand too much, because if the pull of the rubber motor pulls it down any, you will be getting more down thrust than was built into the plastic bearing, and your model will not want to climb because the prop is pulling the front down too much.

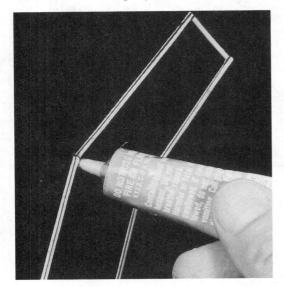

Adding a little extra glue skin after the wing has dried well is good insurance that the tip will stay put. Recheck the dihedral angle in a few minutes to make sure your glue skin did not soften the original glue and allow the angle to change.

Bend up the ends of the undercarriage legs to hold the wheels on.

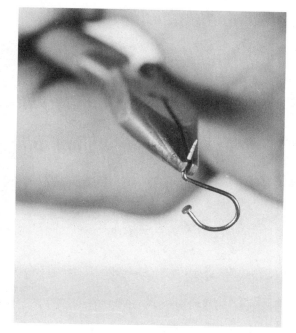

Bend up a rear motor pin as shown. Both round-nose or needle-nose pliers will work.

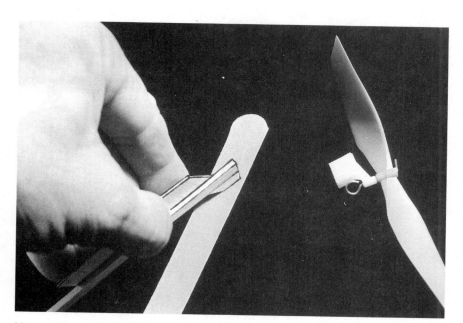

Using an emery board (the one your mom uses for fingernails), carefully take off just enough stock on the spacer block to let the propeller assembly be pushed on snugly. Don't take too much, or a fully wound rubber motor could cock the assembly, giving you more downthrust than is supposed to be there.

Bend rear pin with thread, and glue both pin and thread.

At this time, add the rear motor pin you have bent as shown in the photo and "tack-glue" the model together with little dabs of cellulose (green tube) glue to see what the model looks like. Take back apart by using a little acetone on a brush at the "tacks."

Chapter 5

Finishing the Peck R.O.G.

I had a kid once in my junior high model club who insisted on trying to glide his models before he covered them. They did not so much glide as plummet. Another young man in my Science Workshop classes took his Piper Cub home to cover it and lost the tissue, so he covered it with toilet paper! You can imagine how that looked! I have had kids cover models with the plastic wrap that was supposed to protect the table (actually, it would work if you used contact cement, but they didn't). I have had kids bring in models with so much model dope (a liquid plastic modelers use to paint on tissue) on them that they looked more like cartoons all shriveled up than flying models. You might not be able to judge a book by its cover, but you can sure tell a model by its covering!

PRE-SHRINKING TISSUE

There are lots of ways to shrink tissue up before it gets on the model. If you don't shrink it first, it will find a way to shrink itself later when it picks up moisture and then does what tissue loves to do—tighten up. One guy I know tapes his tissue on the window pane and sprays it with water. Another wads it up into a tiny ball and then spreads it out and irons it to give it an "alligator-skin" look which allows for some accidental shrinkage after it's on the model. One guy out in Oklahoma sprays his with a water mist and then irons it between two sheets of newspaper until it's dry. All of these methods work, and you can take your pick.

   The method I personally prefer is to cut a hole about a half-an-inch smaller than the piece of tissue I have, out of the side of a good-sized cardboard box. Then I glue or tape the tissue over the box. Using an old hairspray bottle filled with water, I spray it until it's good and wet, and then

Secret Number One for helping you do a better job of covering your model: always use *model* tissue.

set it aside to dry. Use a hair dryer on it at your own risk, because tissue has almost no strength at all when it's wet, and you can easily blow a hole right through it. I usually use a dryer after it's already dried out, just to make sure.

When you are ready to use the tissue, cut it out of the hole. If you live in an area where it is damp, or where the model might get wet on damp grass, you might want to use spray lacquer like Testors Dull-Kote or thinned model airplane dope (mix dope thinner with it half-and-half). Any dope will do, but clear is the lightest.

There is nitrate dope, tautening and non-tautening brands, as well as hot-fuel-proof dope on the market. Old timers prefer the tautening nitrate dope for most models, even though it can warp thin wings and tail parts. It has one advantage, and that is that it does its shrinking and gets it over with, and some models can stand the tightening effects easily. Because we're not doping the tissue on the model but on a pre-shrink box-frame, who cares how much it shrinks? After giving the tissue a spray or brushing with the thinned dope, let it dry. Then let it dry some more. Ideally, you would let it age a couple of weeks to make sure it had stopped tightening up. Now's when a few minutes with a hot hair dryer can speed up the aging process.

Now that the tissue is ready to use, you can either cut it out of the frame or you can stick the flat parts directly to it while it is stretched. I like to take it out of the frame and put it on the model. It doesn't give as smooth a covering job, but then, I want a slightly imperfect job so it will have a little bit more space to shrink up without twisting my wing. Besides, people will know I did it—if it looks too good, they might think someone built it for me.

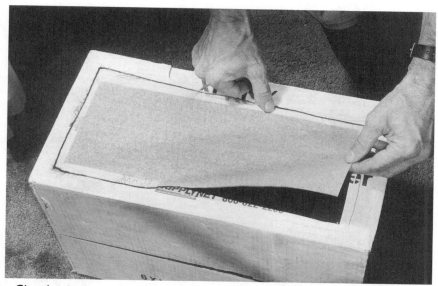

Glue the tissue over a cardboard box with a hole cut in it to pre-shrink the tissue.

Using an old hair-spray bottle filled with water, spray evenly over the tissue.

Modelers use about anything to stick tissue on, ranging from shellac to glue sticks. Two popular ways are to coat the part of the model frame that the tissue will touch about three times with dope and let it dry. Then apply the tissue dry, and brush dope thinner on so that it soaks through the tissue and sticks on it from the bottom up. This is a lot of work, but it gives nice results.

Lazy modelers like me mix about two parts white glue with one part water and brush on one coat. Position the tissue over it carefully and let it down onto the frame. You then lift it carefully at any corner where there is a wrinkle and gently lay it back down, pulling out the wrinkle.

Brush a mixture of 60% white glue, 40% water, on the top of the frame where the tissue is going to stick.

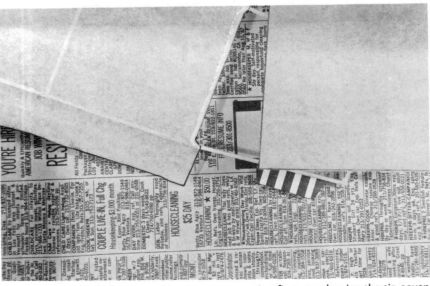

Cover the wing in three pieces. Do the center section first, overlapping the tip covering at the dihedral joint.

Using your thumbs, start at the center and gently apply a little pressure down on the tissue, sticking it onto the wood while very carefully moving your thumbs outward, pulling out little wrinkles as you go. . . The trick here is to keep wiping your thumbs on your fingers every couple of seconds to keep them dry. Once they get sticky, you'll pull chunks out of the tissue as it sticks to you instead of to the wood!

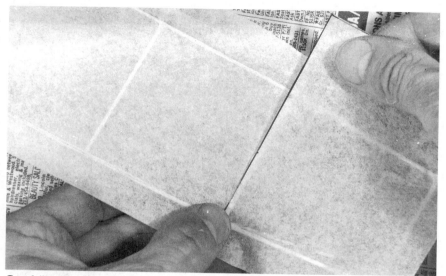

Carefully place the tissue over the area to be covered, then lower it down onto the wet glue.

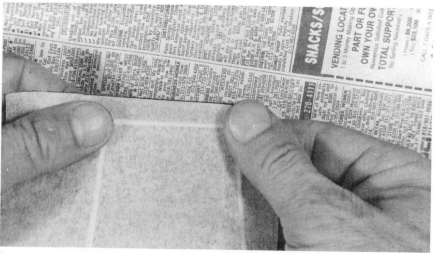

Smooth wrinkles by a gentle, outward action of your thumbs. Keep them wiped dry or the tissue will stick to you instead of to the model.

If you like to do neat work, or if you are short on tissue, take some newspaper and cut patterns just about 1/4″ larger all around for each piece of covering you need. Leave less on the edges of the center section and none at all on the inside of the two wingtip pieces where they will just slightly overlap the center piece. When using the pattern to cut the tissue, pay attention to the *grain* of the tissue, the direction in which it tears easiest. Try a corner first. The grain should run from tip to tip on the wing or *spanwise*. This will become more important when we begin building wings with *cambered*, curved on top, ribs.

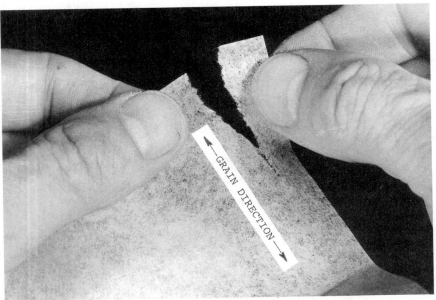

Find the grain direction by tearing a corner of the tissue. It tears easiest in the direction of the grain.

You will have to decide whether or not you wish to cover all your parts and then glue them together (a very neat way to do it) or build the airplane and then tissue it. What I like to do is tack-glue the model together to see how it looks, and then soak the places apart with some acetone on a brush where I tacked it. Then I cover the parts and reassemble them.

Another decision on the Peck R.O.G., which gets covered on one side only of each part to save weight, is whether to cover on the top or the bottom side. Most of us cover on the top because it looks right, but covering on the bottom may give a wee bit more lift, and it also allows the wing to be covered with one piece of tissue all the way across instead of in three parts, which is the way you're going to do the top.

After the glue has dried for about a half hour, it can be trimmed off with a new, sharp razor blade. Use old ones at your own risk; you'll see

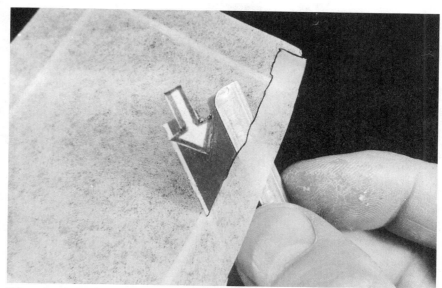

When the tissue dries, trim off the excess with a series of short, downward, shearing motions with a new razor blade.

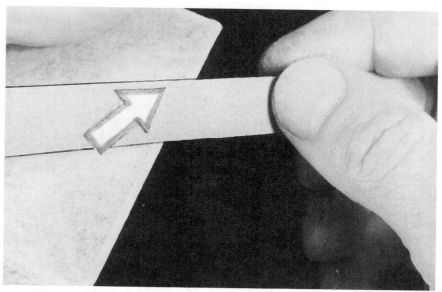

Using an emery board, clean up the edge of the structure after trimming.

why. It takes a gentle touch to move the blade along right up against the LE or TE, holding it at an angle and moving the blade downward with a sort of "sawing" stroke as you go. The idea is to shear it off even with the wood.

Some modelers prefer to use a sanding block and sand through the tissue along the sharp edge of the wood. This can make a very neat job, but be careful not to remove too much of the wood underneath, and don't break the balsa sticks. When you have finished, go along the edge with your finger, wetted with a little white glue, and rub down any rough edge remaining.

If you mess up the covering badly, you can always soak the whole thing in water to get it off, as white glue is water soluble. (The cellulose glue you used to build the model is waterproof.) If you have to start again, pin the parts down flat for a couple of days until they are thoroughly dry. Nobody messes up the job a second time! Well, almost nobody.

When you assemble covered parts, it is not a bad idea to scratch away a little of the tissue where the part attaches. Gluing wood to wood is always stronger than gluing wood to tissue! Before everything dries completely, give the model a quick once-over, with one eye closed, at arm's length just to be sure everything is straight. (See chapter 2 for pre-flight instructions.)

ADDING THE ADJUSTMENT TABS

The adjustment tabs on the wings and the tail of the Peck R.O.G. are not just there for decoration. We say the model has a *low* aspect ratio when a wing has a fairly short wingspan for its *chord*, or distance from the LE to the TE. The Sleek Streek we made earlier had a higher aspect ratio and did not tend to roll quite as much on torque. I suggest replacing the tabs cut out of the plan with ones cut from a 3"×5" card because the latter will be harder to knock out of adjustment. The aileron tabs are important to control the roll, keeping the *bank*—dropping of a wingtip—to a minimum.

The elevator tab is essential on this model because the amount of positive incidence or angle of attack built into the wing mount is pretty small. Without some "up" elevator, the model will dive into the dirt and return itself to kit form on the first flight. You can cut down on the

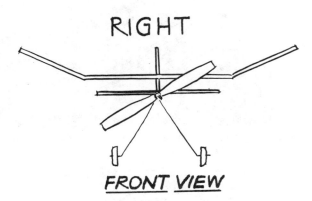

After covering the parts, assemble them as shown with the stab level.

amount of tab needed for elevator if you sand a small angle (negative) under the rear of the fuselage and mount the stab on the bottom. Taking about half the 3/16″ thickness off at the very end and tapering it to full width about 2″ forward should do it. You can use about a third of the original tab size; its purpose will be minor adjustments—not something necessary to make the plane fly.

The finished Peck R.O.G. coming in for a landing, with the last knots of the rubber motor unwinding. Note trim tabs (striped) on wings, stab, and rudder.

POWER

I have had Peck R.O.G.s fly very nicely on 1/16″ FAI rubber indoors in a gym, with times over a minute. The less power you need the better, because flying fast makes any small warps become very effective in ruining your flight. I'd suggest using a 1/16″ motor about 12″ long for your first test flights, even outdoors. You can go up to 3/32″ rubber when you get it flying well.

Remember that the Peck is quite a bit more fragile than your Sleek Streek, and testing under fewer turns in the motor is a good idea, slowly building up to full winds and bigger motors. Indoors, the idea is to make a motor just long enough so that the model touches down just as the last couple of turns run out. If the model's prop stops near the ceiling, the motor is too short or too powerful. If it lands with winds left, it is too long or too weak.

Sometime you might want to try adding a little more twist into the blades of the prop so that it will turn a bit slower indoors and give you a longer motor run. Outdoors, you'll probably want the extra prop speed for a high climb. I suggest you take this book or a photocopy of the flying instructions and troubleshooting chart to the field or gym with you.

REPAIRS

Although you will do your best to avoid damage, things always seem to get broken on models—either while building, on the way to and from the flying site, or (once in a great while) in a crash. A good thing to remember is that your model is never as badly off as it looks!

I have had kids in my classes throw perfectly good models away with only five or six breaks. Remember how many places needed to be glued together when you started making it? Well, it probably needs fewer joints now!

The main thing is to try and get as much gluing surface as you can. Gluing two sticks together end-to-end is never as strong as sanding or cutting an angle on each one and gluing the longer surface together. A 1/4" fuselage "butt-jointed" end-to-end gives 1/4" of gluing surface (for purposes of illustration), whereas if you cut the wood at 45°, you'd increase the gluing area to 1/2". The more gentle the angle, the more surface. Of course, you need to put an extra piece into the joint (see fuselage break illustration) or you will be too short! Sometimes, if you're lucky, the break will cover a fairly large area, and, when it is glued back together it will probably be stronger than the original place where it broke. Always double-glue.

A quick field repair on a broken model can often be made by adding a splint made from scrap balsa to the repaired joint. This adds strength and can help get the piece straight again. An extra thickness of balsa is often called a *doubler* and is sometimes added in places where breaks might be expected before they happen: where the wing LE and TE attach to the pylon wing mount, for example.

After gluing a broken section, it is a good idea to pin it down to something flat, or to put some weight on it. If you use the corner of your model box top for this, and have no plastic wrap handy to keep the glue from sticking, you had better make sure all extra glue is squeezed out first and wiped off so the model will not become a permanent part of the model box. Pieces of balsa sheet, if they are flat, often make good devices to pin wing tips and such to while the model is sitting on its wheels. This way, two or three repairs can be drying at the same time.

Leading edges of wings often get damaged, as they hit first. Add a doubler in behind the break, so it will be hard to see and out of the airflow (to keep drag down). Cut the ends of the repair doubler at angles or taper it toward each end from the middle to make it lighter and a bit more flexible.

After making repairs and before the glue dries completely, line up the parts to make sure the fuselage is straight and the wings are unwarped.

"GOOD" BREAK

JUST GLUE BACK TOGETHER
(DOUBLE GLUE)

"BAD" BREAK

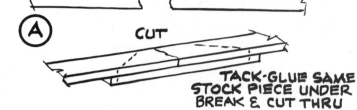

(A) CUT

TACK-GLUE SAME
STOCK PIECE UNDER
BREAK & CUT THRU

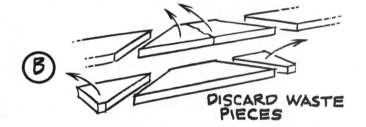

(B)

DISCARD WASTE
PIECES

(C) REASSEMBLE NEW
COMPONENTS IN STRAIGHT LINE
(OVER SARAN) AND DOUBLE-GLUE

OR...

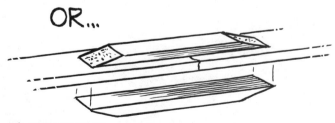

ADD DOUBLERS EACH SIDE
CHECK FOR STRAIGHTNESS WHILE DRYING

How to repair a broken fuselage.

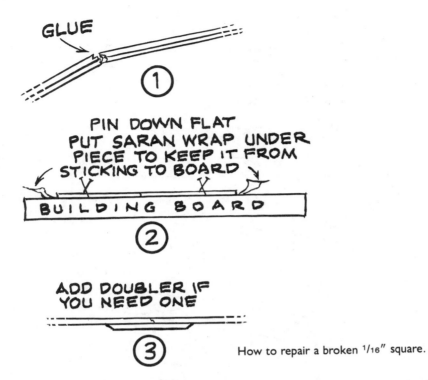

How to repair a broken 1/16″ square.

Move the parts into alignment as many times as necessary to get it right.

You can repair tears in tissue with a little line of cellulose cement (our trusty old Testor's green tube or Ambroid) along the tear line if it matches up pretty well—not too much, as it will shrink a little. You can use frosted "Magic Mending" tape in a pinch.

The ideal repair would be to cut the injured section out all around the balsa frame and make a patch to cover the whole area. You can use the square you cut out as a pattern for the new tissue, cutting it a little bigger so there will be something to glue to, and just re-cover that part. If you only tissue-patch a part of the area between the ribs or between the LE and TE, it will stand out like a sore thumb.

A final note on repair concerns added weight. If you added weight behind the wing with your repairs, you will probably need to add some weight to the nose with a little modeling clay to move the CG back where it belongs—unless you were nose-heavy to start with! A heavier, repaired right wing might force you to add a little weight to the left wing, unless you need some right turn.

DEALING WITH WARPS

Thin wings and tail surfaces will warp or twist until they look like potato chips at the least excuse. The main causes are weak balsa structure and tissue that shrinks when it gets wet or damp and then dries out. Adding

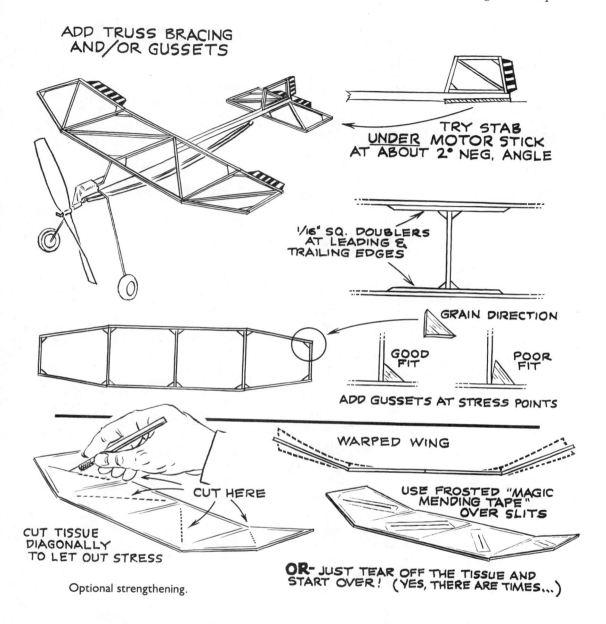

ADD TRUSS BRACING AND/OR GUSSETS

TRY STAB UNDER MOTOR STICK AT ABOUT 2° NEG. ANGLE

1/16" SQ. DOUBLERS AT LEADING & TRAILING EDGES

GRAIN DIRECTION

GOOD FIT

POOR FIT

ADD GUSSETS AT STRESS POINTS

CUT HERE

CUT TISSUE DIAGONALLY TO LET OUT STRESS

WARPED WING

USE FROSTED "MAGIC MENDING TAPE" OVER SLITS

OR- JUST TEAR OFF THE TISSUE AND START OVER! (YES, THERE ARE TIMES...)

Optional strengthening.

gussets at the corners of wing and tail structure as shown in the drawing helps. A *gusset* can be a little triangle of balsa glued in a corner, or it can be a 1/16" square bit of balsa with the ends sanded at little angles about the same length as the wide end of a triangular gusset. Angle braces or *truss* bracing can be added from corner-to-corner across the wing *bays*, as shown. A strong wing-to-pylon joint will help, too—as when one side of the LE warps "up," the opposite side warps "down." The heavier the grade of balsa sticks used for LE and TE, the less warping you will get. All these additions make the plane heavier and cut down on flight times. Then again, a badly warped wing makes flight *impossible*.

Using pre-shrunk or pre-shrunk-and-pre-doped tissue is a big help. Even so, the weather can play dirty tricks on you. I live near the beach, and anything I cover is a tiny bit damp. Tonight, the wind is blowing from the desert, and I can tell how dry it is by looking at my Peck R.O.G.s. "Pringle City!" I could avoid this by covering models out in the desert, or at least waiting for the driest, sunniest day I can find to cover. Some indoor model builders even make cardboard boxes with electric light bulbs in them, and cover their models inside, making sure no damp air can get in by sticking their arms through holes cut in the box while they cover!

On the flying field, most modelers try to take warps out by breathing moisture on the wing and then twisting it in the opposite direction. If that does not work, you might try to crack the LE or the TE and reglue it without the warp. Then if that does not work, try to overpower the warp with a control tab made of card stock or balsa glued to the TE and bent opposite the warp to try and cancel it out. The next step might be to try slitting the tissue (as shown in the illustration) and then, when the wing is flat again, covering the slits with tape. Really bad warps might require removal of the warped part from the plane and soaking it uncovered in water overnight. Then pin it down flat for a few days, or cook it in the oven while it is pinned to a flat surface at LOW heat for an hour. Then re-cover it with pre-shrunk tissue. The worst thing that can happen is to have to make a new wing, which only takes about ten minutes.

WARREN SHIPP vs. THE PECK R.O.G.

Warren Shipp used to be a guard in the New York Subway. Because his was a pretty boring job—arresting muggers, critiquing spray-can art, and the like—he used to dream of retiring and building model airplanes. When he did retire, he turned to crazy stuff like autogiros—probably the result of too much carbon monoxide down in the tubes or something.

When Warren saw the Peck R.O.G., something in his mind snapped, and he began a campaign to drive Bob Peck nuts, too. Each time Bob would see him, he would have a new, modified version of Peck's wee wonder. One time it would be a canard (tail-firster), the next time it would be a biplane, etc, etc. You might want to try some fun things with it yourself, but I warn you, Warren has probably beaten you to it.

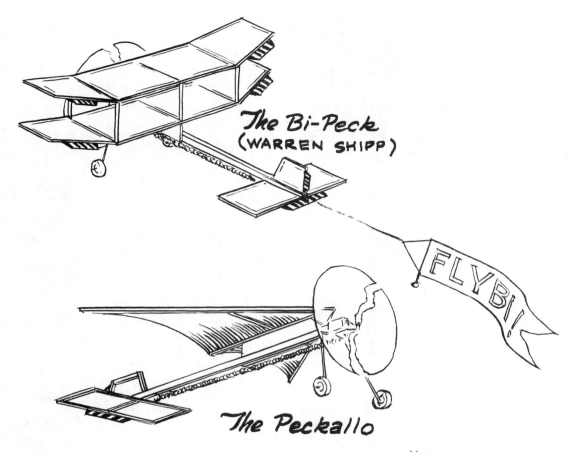

The Bi-Peck
(WARREN SHIPP)

FLYB!

The Peckallo

Variations of the Peck R.O.G.

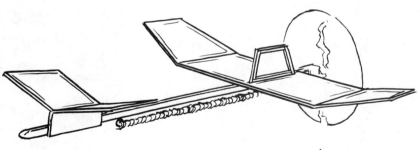

The PushPeck

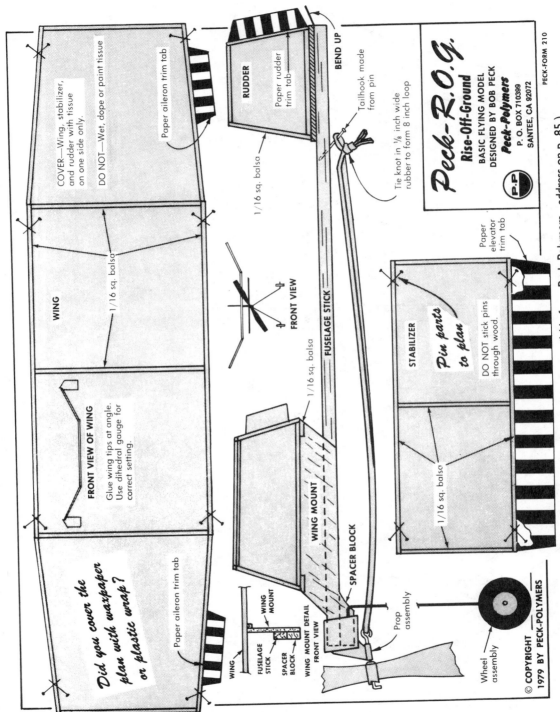

COVER—Wing, stabilizer, and rudder with tissue on one side only.

DO NOT—Wet, dope or paint tissue

Paper aileron trim tab

RUDDER

Paper rudder trim tab

BEND UP

Tailhook made from pin

Tie knot in ⅛ inch wide rubber to form 8 inch loop

Peck-R.O.G.

Rise-Off-Ground

BASIC FLYING MODEL
DESIGNED BY BOB PECK

Peck-Polymers

P. O. BOX 710399
SANTEE, CA 92072

PECK-FORM 210

1/16 sq. balsa

WING

1/16 sq. balsa

FRONT VIEW

FRONT VIEW OF WING

Glue wing tips at angle. Use dihedral gauge for correct setting.

FUSELAGE STICK

1/16 sq. balsa

Paper elevator trim tab

STABILIZER

Pin parts to plan

DO NOT stick pins through wood.

1/16 sq. balsa

Did you cover the plan with waxpaper or plastic wrap?

Paper aileron trim tab

WING MOUNT

SPACER BLOCK

WING

FUSELAGE STICK

WING MOUNT

SPACER BLOCK

WING MOUNT DETAIL FRONT VIEW

Prop assembly

Wheel assembly

© COPYRIGHT
1979 BY PECK-POLYMERS

Reduced plans of the Peck R.O.G. (Full-size plans available from Peck Polymers—address on p. 85.)

Appendix

Suppliers & Publications

Everything covered in this book is available from Peck Polymers as a convenience for modelers living away from good sources of modeling materials. Also listed are other suppliers who carry the same types of useful kits, plans, and supplies. Always send an SASE (self-addressed, stamped envelope) when requesting information, as many of these sources operate on a shoestring! Sending cash through the mails is not advised; a check or postal money order is better. A more complete list will be included with the other books in this series, where more advanced projects are covered.

A.A. Lidberg Plans
614 E. Fordham
Tempe, AZ 85283

Lidberg has many "profile" model plans that are tissue-covered and easy to build. He has more advanced plans, too, with excellent instruction sheets.

Blue Ridge Model Products
880 Carmen Ct.
La Verne, CA 91750

Good model kits for beginners with some of the work done for you. Wood is excellent. One of my students got a nine-minute thermal flight with one of their beginners' hand-launched glider kit models. Small fee for a catalog.

Easy Built Models
Box 1059
Beamsville, Ontario, LOR 1B0
CANADA

Large selection of model kits, plans and supplies from the "good ol' days." Prices are reasonable, and the kits range from quite simple to complex. Some of the designs are a bit primitive by today's standards, and the wood might not always be what you would wish to use, but despite this, they have some interesting stuff. You can get their list of kits and an interesting booklet called *Facts and Building Tips* for a small charge.

FAI Model Supply

P.O. Box 3967
Torrance, CA 90610

Rubber by the pound, special rubber lube, some more advanced kits. Fast, reliable service. Catalog available for a small charge.

Gene DuBois Models

P.O. Box "C"
Acushnet, MA 02743

Small but nice range of easy to more difficult kits. Small charge for a catalog.

Indoor Model Supply

Box 5311
Salem, OR 97304

Lew Gitlow carries lots of stuff for indoor modelers plus some tested beginners' models like the Yard Bird, Slow Poke, and 12″ hand-launched gliders. Wood selection for the kits is outstanding.

Mace Model Aircraft Co.

359 So. 119th. East Av.
Tulsa, OK 74128

Don Mace has a couple of beginners' indoor fliers called the P-18 and P-24 (number = inches of wingspan). Great fliers, and easy to build. Send large SASE for listings.

Midwest Products Co., Inc.

400 So. Indiana St.
P.O. Box 564
Hobart, IN 46342

Teachers, Scout, and Youth Group leaders will be interested in Midwest's model programs. They sell class sets of beginners' models, including the Frank Zaic-designed X-18, which was used for many years as the backbone of my Space Museum classes and which is one of the easiest and best-flying models ever sold. You can get a sampler of their five project models (set #520) by sending $8.95. **Note**: Adding an extra thickness of scrap balsa shim between the leading edge and fuselage stick of the X-18 generally improves its performance.

Old-Timer Model Supply

P.O. Box 7334
Van Nuys, CA 91409

Ken Sykora's "1930's Model Shop" is just that. No RC cars or indian beadcraft—just stuff like plans, balsa, nitrate dope, tissue, balsa prop blanks, rubber, and an exciting catalog—an instant collectors' item—which you can get for $2.00.

Peck Polymers

Box 710399
Santee, CA 92072
Telephone: (619) 448-1818   FAX (619) 448-1833

This is the one-stop source of the stuff used in this book. It is highly recommended. Model-for-model, Peck has more winners than any other manufacturer I know. The model projects covered in Volumes 2 and 3 of this series also come from Peck's. Their line of really good kits, supplies, and plans is second to none for the beginning modeler, though you would probably not be interested in their RC blimps just yet! Their extensive and fascinating catalog is $2.00. If you would just like a listing of the models we use, send a large SASE and ask for the "Hey Kid!" list for free.

SIG Mfg. Co.

401-7 South Front St.
Montezuma, IA 50171

SIG is big. Despite an emphasis on the higher-profit RC items, they do carry a number of beginners' kits, among which is the SIG Cub 24 (which benefits from using a 9" plastic Peck prop instead of the one supplied). They have winders, rubber, lube, dope, tissue and other useful stuff for beginners. Their huge catalog is $3.00.

Simple Simon Airplane Co.

P.O. Box 18
East Longmeadow, MA 01028
Phone: (413) 592-3615 or (413) 782-3872

Simple stick-fuselage model kit for beginners (easier than Peck R.O.G.) and a four-pack of all-sheet variety gliders for youth groups. A case of 24 Simple Simon 14" wingspan ships is $59.76 plus shipping and comes with a money-back guarantee! Step-by-step building instructions.

Superior Aircraft Materials

12020-G Centralia
Hawaiian Gardens, CA 90716
Phone (213) 865-3220

Good source for bass and spruce as well as quality balsa wood, though it will cost you about $25, due to minimum order requirements and handling + UPS. The owners are top modelers and can handle special requests.

Model Aviation

1810 Samuel Morse Dr.
Reston, VA 22090

Official publication of the Academy of Model Aeronautics which promotes modeling and sanctions contests in the US. Covers all aspects of aeromodeling. Subscription is $18/year regular, $13.50 for libraries, and $9.00 for schools. Membership in AMA includes the magazine and provides the member with insurance while flying models or on the way to contests. Membership rate is $40 for adults, with lower rates for seniors and young modelers.

Model Builder Magazine

898 W. 16th St.
Newport Beach, CA 92663

This is the magazine in which the material in this book first appeared. Covering all aspects of modeling, including even model boats, *Model Builder* has more free-flight space than any other magazine, ranging from beginners' models to old timers and scale. Editor Bill Northrop has for a long time been committed to promoting free flight as the base from which all other aspects of the hobby grew. Subscription $25/year.

Flying Models Magazine

P.O. Box 700
Newton, N.J. 07860

Flying Models usually has material of interest to free-flighters, including model construction articles by leading free-flighters and columns by knowledgeable and active modelers. Subscriptions $25/year.

BEGINNERS' MODEL BOOKS

Peck Polymers (see address in the appendix)

Peck has several books in stock, including this one, one hopes! *Rubber-Powered Model Airplanes* by Don Ross is one you might find useful.

Hannan's Runway

Box 210
Magalia, CA 95954

Hannan's carries the Ross book and a book on indoor scale models by Fred Hall, *Throw It Out of Sight*, about hand-launched gliders. Also available are British model books with old-time models in them, and the Zaic Yearbooks covering many types of historic and interesting models and subjects. Send $1.00 for latest listings.

Index